ATLAS OF BENTHIC SHELF FORAMINIFERA OF THE SOUTHWEST ATLANTIC

ATLAS OF BENTHIC SHELF FORAMINIFERA
OF THE
SOUTHWEST ATLANTIC

Esteban Boltovskoy / Graciela Giussani
Silvia Watanabe

Museo Argentino de Ciencias Naturales and Consejo Nacional de Investigaciones Científicas y Técnicas, Avda. Angel Gallardo 470, 1405 Buenos Aires, Argentina

Ramil Wright

Department of Geology, Florida State University, Tallahassee, FL 32306

Dr W. Junk bv Publishers The Hague–Boston–London 1980

Distributors:

for the United States and Canada

Kluwer Boston, Inc
160 Old Derby Street
Hingham, MA 02043
USA

for all other countries

Kluwer Academic Publishers Group
Distribution Center
P.O. Box 322
3300 AH Dordrecht
The Netherlands

Library of Congress Cataloging in Publication Data   CIP

Main entry under title:

Atlas of benthic shelf foraminifera of the southwest Atlantic.

Bibliography
  1. Foraminifera – South Atlantic Ocean.   2. Benthos – South Atlantic Ocean.   3. Protozoa – South Atlantic Ocean.   I. Boltovskoy, Esteban.
  II. Giussani, Graciela.   III. Watanabe, Silvia.   IV. Wright, Ramil.
QL368.F6A74   593.1'2098   80-239
ISBN-13: 978-94-009-9190-3   e-ISBN-13: 978-94-009-9188-0
DOI: 10.1007/978-94-009-9188-0

TABLE OF CONTENTS

INTRODUCTION

Benthic foraminifera from the southwestern Atlantic have been studied since 1839. However, despite the appearance of about 60 articles dealing with the benthic foraminiferal fauna of this area, there is no single work which has attempted to synthesize the taxonomy, distribution, and ecology of the fauna. Many of the published papers deal with portions of the area, and one is even a summary of the zoogeography and ecology of South America (Boltovskoy, 1976). It is one purpose of this work to bring together in one place the descriptions and illustrations to accompany and amplify the zoogeographic and ecologic work done in the past.

The nomenclature of benthic foraminifera has undergone many changes since 1839 when d'Orbigny first described forms from South American coastal waters. In addition to gradual changes in ideas about foraminiferal taxonomy, there have been significant technical changes in both optical and electron microscopy which have greatly altered our notions about the construction and ornamentation of the foraminiferal test. Details of microarchitecture and crystallographic orientation have given us information unavailable to past observers, and consequently have modified our views of classification. A restudy of existing data and material is clearly in order and it is an additional purpose of this work to revise and update the taxonomy given in past studies in the area.

To accomplish these goals we utilized and studied material from a variety of sources, of which the principal is the collection of the Laboratorio de Foraminíferos, Museo Argentino de Ciencias Naturales "B. Rivadavia" in Buenos Aires. This collection contains approximately 10,000 foraminiferal slides consisting primarily of material from the SW Atlantic supplemented by several hundred comparison slides sent to the laboratory by colleagues around the world. In addition, special collections were made from those parts of the Argentine zoogeographic province whose benthic foraminiferal fauna was poorly known. To supplement the material housed in the Buenos Aires collection, the senior author visited various institutions to study the original material of investigators such as d'Orbigny, Williamson, Brady, Cushman, and Heron-Allen and Earland. Due to space limitations we have figured and described only those species which are important by virtue of their abundance, widespread lateral extent or restriction to a single subprovince or environment.

The majority of the samples on which this study is based were not preserved at the time of their collection. Consequently, it was not possible in many cases to distinguish between living and dead tests. For this reason, the data on the distribution of the various species is essentially based on the occurrence of empty tests. However, we believe that the qualitative distribution of empty tests on the Argentine shelf is representative of the distribution of living forms.

The bulk of the material studied was collected during numerous cruises sponsored by the Argentine Hydrographic Service, whom we wish to acknowledge and especially thank for their constant cooperation over a period of many years. Their devotion to the study of the biology of Argentine waters is greatly appreciated. The remaining samples were collected principally by the senior author and various colleagues and assistants who have been mentioned in earlier publications. We thank Marly Madeira-Falcetta for supplementary material from Brazilian waters. We are also glad to acknowledge the assistance of the Scanning Electron Microscope Service of the Argentine National Council of Scientific and Technical Investigations in the preparation of the photomicrographs included in this work.

AREA OF STUDY

This study is based on those benthic foraminifera of the southwest Atlantic which occupy the broad continental shelf off the Argentine, Uruguayan and south Brazilian coasts.

A zoogeographic province is an area characterized by a fauna, in this case benthic foraminifera, which is distinct from those of adjacent zones. The differences between provinces are due primarily to differences in latitude, the geological evolution of the area, and water circulation patterns. Along the continental borderlands of the southwest Atlantic the latitudinal situation is rather straight forward: the coastal margin as well as the shelf edge roughly follow lines of longitude. Consequently, latitudinal control over faunal distributions is marked. The current latitudinal control has been in effect since the closing of the Isthmus of Panama eliminated tropical water interchange between Atlantic and Pacific about 3.5-4.0 m.y. ago. The shallow water circulation patterns in the southwest Atlantic began to take on their present configuration in the late Cretaceous ($\sim$90 m.y.) with the final separation of Africa and South America. The continued opening of the South Atlantic during the Eocene created gaps in the Rio Grande and Walvis ridge systems with subsequent initiation of deep water circulation about 35 m.y. ago (Berggren & Hollister, 1974). The opening of the Drake Passage during the late Cenozoic set the stage for circum-antarctic flow and the potential for immigration of faunas from the Pacific area.

The specific water mass circulation patterns in the area are a significant factor in explaining the foraminiferal distribution. In order to better understand the distribution, character and divisions of these influencial water masses, we present below a brief discussion of the hydrologic scheme of the southwest Atlantic (Fig. 1). This description is based in large part on the work of Boltovskoy (1970b).

WATER MASSES

Cape Horn Current. The Cape Horn Current is the northernmost part of the circumpolar current or West Wind Drift. It consists predominantly of cold subantarctic water with a slight influence from the warmer waters of the South Pacific. This influence is evident in the surface waters south of Tierra del Fuego where the water temperature is higher than in the surface waters more to the east where there is no Pacific influence. The Cape Horn Current, after passing south of Tierra del Fuego, migrates northeast where it loses its identity by becoming mixed with waters of the West Wind Drift. A branch of the Cape Horn Current moves northward to form the Malvin Current.

Malvin Current. The Malvin Current is initiated from a western branch of the Cape Horn Current at a point beyond the influences of waters of Pacific origin. The Malvin Current, like the Cape Horn Current, is composed of subantarctic water. Its surface temperature in the south (46-50°S) is as low as 5-7°C in winter and in the north (40-44°S) is as high as 10-11°C in summer. Its salinity commonly fluctuates between 33.5-34‰. The Malvin Current is rather strong and can reach velocities as great as 0.6-0.8 knots. The eastern limit of the current is located beyond the edge of the continental shelf and consequently is not discussed here. Using benthic foraminifera as hydrologic indicators its western boundary along the substrate was located between 36°30′S and 42°00′S (Boltovskoy, 1959a, b). Its western limit at the surface was defined on the basis of planktic foraminiferal studies (Boltovskoy, 1970b). The nearshore boundaries on the bottom were extended as far south as 52°S in a study by Giussani & Watanabe (in press). The western limits of this current are shown on Table 1. From Table 1 and Fig. 1 it is evident that the western boundary of the Malvin Current generally bends westward toward the continent as it passes downward from the surface into the bottom waters. In a very general sense the western boundary of the current coincides with the 80-100m isobath.

The Malvin Current continues northward at the surface as far as 35°30′-36°30′S where its waters sink and become partially mixed with water of a different origin. Nevertheless, the water retains some of its identity as it sinks below the warm waters of the Brazilian Current. The cold water continues to move over the sea floor as far northward as 21-22°S where all vestiges of its identity are lost.

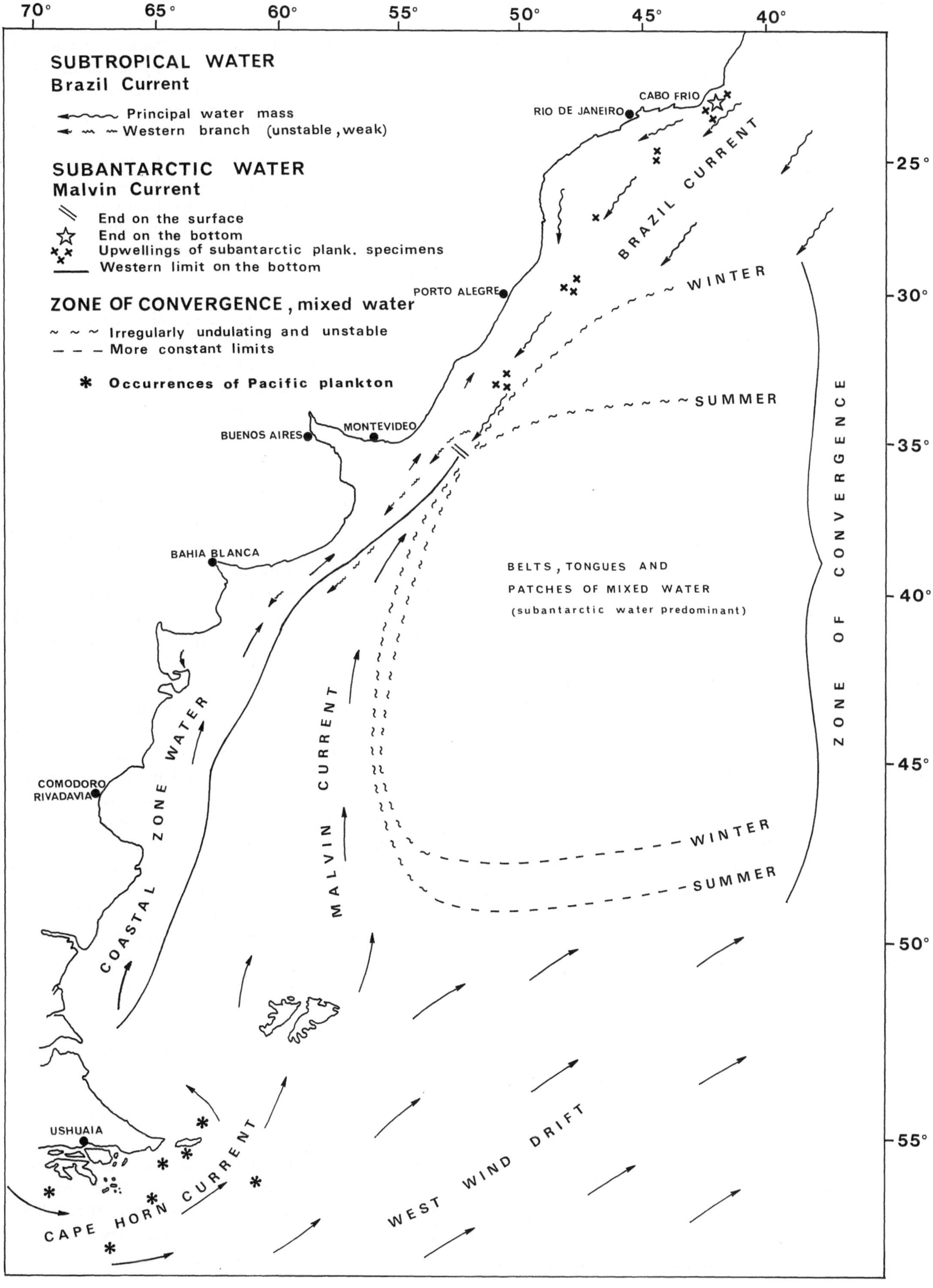

SUBTROPICAL WATER
Brazil Current

⟶∿ Principal water mass
◄∿ Western branch (unstable, weak)

SUBANTARCTIC WATER
Malvin Current

⫽ End on the surface
☆ End on the bottom
ˣˣ Upwellings of subantarctic plank. specimens
— Western limit on the bottom

ZONE OF CONVERGENCE, mixed water

∿∿∿ Irregularly undulating and unstable
— — — More constant limits

✶ Occurrences of Pacific plankton

RIO DE JANEIRO
CABO FRIO
BRAZIL CURRENT
WINTER
SUMMER
PORTO ALEGRE
BUENOS AIRES
MONTEVIDEO
BELTS, TONGUES AND
PATCHES OF MIXED WATER
(subantarctic water predominant)
BAHIA BLANCA
COASTAL ZONE WATER
MALVIN CURRENT
COMODORO RIVADAVIA
WINTER
SUMMER
ZONE OF CONVERGENCE
USHUAIA
CAPE HORN CURRENT
WEST WIND DRIFT

Fig. 1.   Surface currents and water masses in the southwest Atlantic (after Boltovskoy, 1970b)

Table 1. Western limit of Malvin Current based on foraminifera. Bottom data from Boltovskoy (1959b, 1970b), and Giussani and Watanabe (in press). Surface data from Boltovskoy (1970b).

	Bottom	Surface	
Latitude, °S	Longitude (W)	Longitude (W)	
		Summer	Winter
35	53°05′	—	52°30′
36	53°05′	53°45′	53°30′
37	54°15′	54°30′	54°45′
38	56°00′	55°15′	56°00′
39	57°05′	56°15′	57°15′
40	59°00′	57°30′	58°45′
41	59°55′	58°15′	60°00′
42	60°25′		
43	61°10′		
44	62°00′		
45	62°45′		
46	62°50′		
47	63°00′		
48	63°30′		
49	64°00′		
50	64°40′		
51	65°25′		
52	66°20′		

Coastal Argentine waters. The Argentine coastal water mass lies between the coast to the west, the Malvin Current to the east, the zone of influence of the Río de la Plata to the north, and although less well studied to the south, appears to end near the mouth of the Río Gallegos (∼52°S). The coastal waters are of subantarctic origin but are mixed with fresh waters from the continent and in the northern part with water from the Brazil Current. The coastal waters generally move from south to north, although local movements are easily complicated by tides and winds.

The surface temperature of the coastal water mass exhibits considerable geographic and seasonal variation (Table 2).

Table 2. Seasonal average temperature, surface coastal water (after Boltovskoy, 1970b).

Latitude, °S	Temperature, °C			
	Summer	Fall	Winter	Spring
32	23.0	19.0	14.0	19.0
35	22.0	16.0	11.5	17.0
38	20.0	14.0	10.0	15.0
41	18.0	13.5	9.0	14.0
44	16.0	13.0	8.0	12.0
47	13.5	10.0	7.0	9.0
50	12.0	8.0	6.0	8.0
53	11.0	7.0	4.5	7.0
56	8.5	5.0	4.5	5.5

Area influenced by the Río de la Plata. The water disgourged into the southwest Atlantic by the Río de la Plata constitutes another biologically distinct water mass. Its limits to the east are defined by the first appearance of planktic foraminifera in the surface waters. This boundary coincides, in a very general sense, with the 30‰ isohaline. The position of this eastern boundary is difficult to define geographically because it changes greatly both in space and time. Its position is changed by the winds, the surface water currents and the rainfall activity in the basin of the Río Paraná.

The waters influenced by the Río de la Plata remain at the surface near their eastern edge and override the Argentine coastal waters. During those times when the Río de la Plata water mass is shifted westward its eastern boundary coincides with the western border of the subtropical water of the Brazil Current. This abutment has the effect of introducing Brazilian faunal elements into the northern part of the Argentine coastal waters.

Coastal waters of Uruguay and southern Brazil. Another set of coastal waters lies in a narrow zone between the continent and the Brazil Current. Their southern limit is the zone influenced by the Río de la Plata. The eastern limit of these waters at the surface is the Brazil Current, and at depth, although not well defined, seems to be the western edge of the diluted Malvin Current which lies below the Brazil Current.

The surface temperature of these coastal waters is a function of latitude. The winter monthly average in the south off the Uruguayan coast is 12°C whereas the summer monthly average near Rio de Janeiro is 27°C. Although the salinity is a bit less than that of the normal open ocean, the waters are certainly euhaline.

The movements of these waters are not well defined. In summer the predominant movement seems to be towards the south-southwest whereas in winter the direction is reversed. These winter movements toward the north-northeast have been observed as far north as 29°-30°S.

Beyond the shelf edge the warm subtropical Brazil Current, moving towards the south-southwest, encounters the cold subantarctic Malvin Current and the West Wind Drift, moving in the opposite direction, to form the subtropical-subantarctic convergence. This zone and its fauna is located in the open ocean beyond the scope of this work.

BENTHIC FORAMINIFERA PROVINCES

The distribution of water masses controls the zoogeography of an area and this influence will be seen in the discussion of provinces and subprovinces that follows. The discussion of the zoogeographic division of South

America on the basis of benthic foraminifera is extracted from the work of Boltovskoy (1976).

Four provinces can be recognized around the South American continent. They are the West Indian, Argentine, Chilean-Peruvian, and Panamanian. Each can be divided into subprovinces. The benthic foraminifera of the Argentine Province provide the main theme for this book. This province (Fig. 2) lies off the southeastern part of the continent, south of the West Indian Province. The boundary between these two provinces lies at abut 32°C and is clearly marked by faunal changes. The southern limit of the province has been less well studied and is not as clearly defined. It lies generally to the south of Tierra del Fuego where the Straits of Magallen enter the Pacific. Beyond this area the benthic foraminifera are characteristic of the Chilean-Peruvian Province.

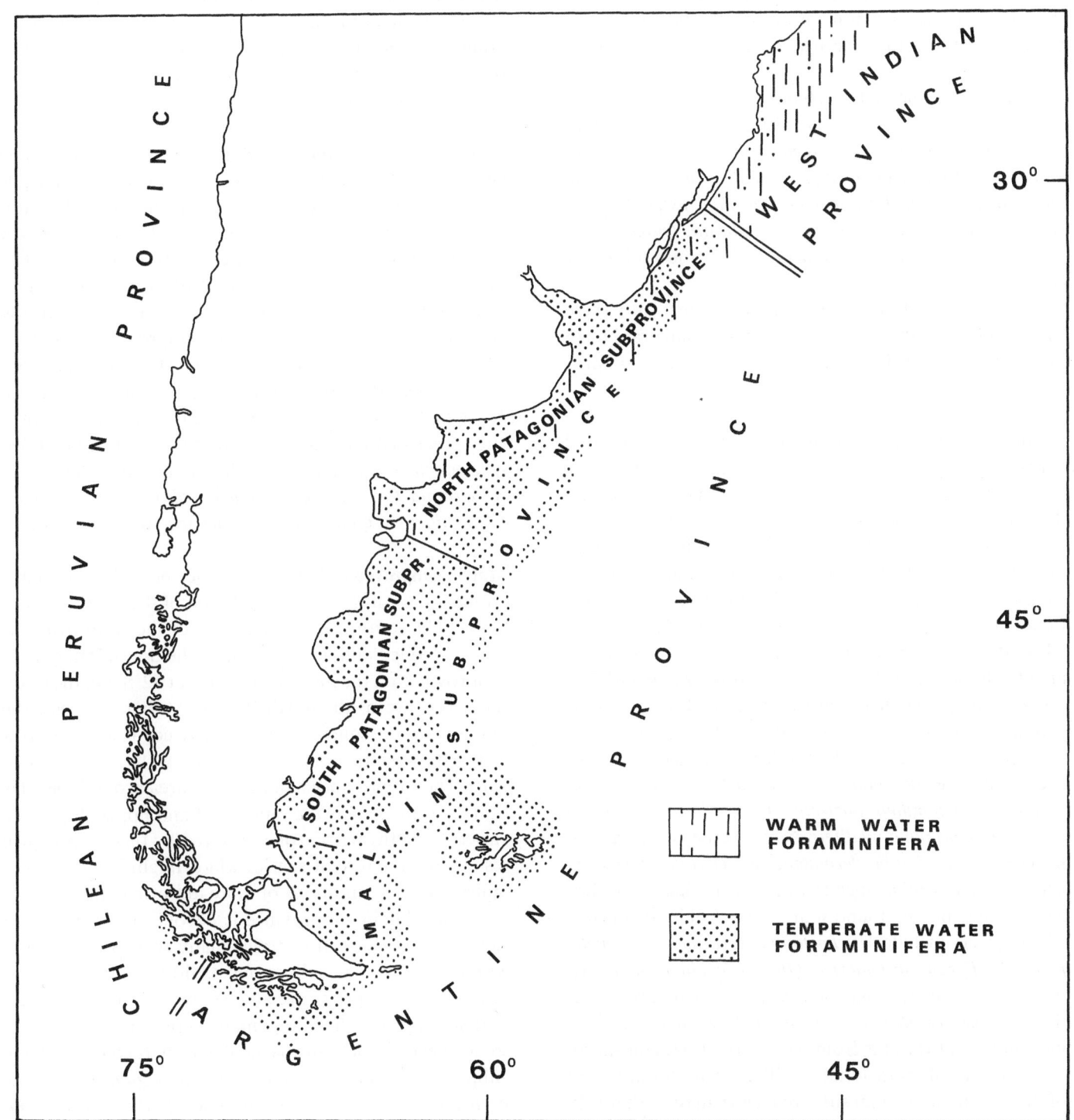

Fig. 2.   Zoogeographic divisions of the southwest Atlantic (after Boltovskoy, 1976)

The Argentine Province consists of temperate waters and is dominated by the species *Buccella peruviana*, s.l. About 650 taxa have been described from the province. This seems to be a rather large number for a temperate zone, but may be an artifact of the taxonomy. The majority of the taxa in this area were cited by Heron-Allen & Earland (1932), authors with a rather narrow species concept. This viewpoint undoubtedly influences the apparent high diversity of the province. We believe that there are many synonyms among the faunal lists for the province and that the true diversity is less than the apparent.

Species unique to the Argentine Province are relatively few (Boltovskoy, 1976) and consist of the following: *Allogromia flexibilis, Ammodiscus plicatus, Asterigerinata pacifica, Astrononion gallowayi, A. stelligerum, Buliminella auricula, Discorbinella altocamerata, Florilus pauperatus, Glabratella chasteri, Notorotalia clathrata, Quinqueloculina arctica, Recurvoides contortus* and various unilocular calcareous species.

The Argentine Province can be divided into three subprovinces on the basis of benthic foraminifera: North Patagonian, South Patagonian and the Malvin subprovinces.

North Patagonian subprovince. This subprovince is located between 32°S and 42°-43°S. Its southern limit generally coincides with the Valdez Peninsula. South of this area lies the South Patagonian subprovince. The distinction between these two subprovinces is based mainly on the distribution of two species of *Elphidium*, a genus quite common to the shallow shelf areas of the Argentine Province. The North Patagonian subprovince is characterized by *E. discoidale* whereas the species *E. macellum* is characteristic of the area to the south of the Valdez Peninsula. *E. discoidale* is a common species in Brazilian waters and it, as well as several others of Brazilian origin, fails to migrate southward past the vicinity of the Valdez Peninsula. The other species include *Ammonia* ex gr. *A. parkinsoniana, E. galvestonense, Globulina australis, G. caribaea, Massilina secans, Planorbulina mediterranensis, Poroeponides lateralis, Quinqueloculina brodermanni,* and *Textularia gramen*. There are a few other species which also originate in Brazilian waters but are found south of the Valdez Peninsula. These species include *Mychostomina revertens, Nonionella pulchella, Pyrgo subsphaerica, Quinqueloculina horrida, Q. intricata, Q. lamarckiana* and *Spirillina vivipara*; however, these species often occur as isolated specimens and as morphological variants from their normal size and shape.

The typical characteristics of the benthic foraminifera of the North Patagonian subprovince seem to be that they are temperate water forms of subantarctic origin, which are mixed with some rare Brazilian elements. This mixing

is due to the presence of a branch of subtropical water west of the Malvin Current. These Brazilian elements may have established themselves in this area during a Miocene episode of warmer temperatures and southward shifting bottom faunas. Another typical trait of this subprovince is that its foraminiferal fauna is somewhat depauperate. The specimens are smaller than normal, often lack ornamentation or exhibit only partial ornamentation, and are sometimes morphologically irregular. Although no satisfactory explanation has been given to explain this phenomenon, it has been suggested that the reasons may lie in the quantity and types of trace elements in the environment (Boltovskoy, 1961).

South Patagonian subprovince. This subprovince is located between the Valdez Peninsula (42-43°S) and Río Gallegos (~52°S). Although the faunal composition of this subprovince is similar in many ways to that of the North Patagonian subprovince it differs in three respects. The foremost is the change from *Elphidium discoidale* to *E. macellum* described above. Moreover, there are practically no Brazilian faunal elements in the subprovince except for a few isolates in the northern part. The final difference lies in the larger and better developed specimens in the south. The depauperate traits exhibited in the northern subprovince are essentially absent to the south. There are a few species which are endemic to this subprovince. They are: *Allogromia flexibilis, Discorbinella altocamerata, Nonionella chiliensis, Quinqueloculina arctica* and *Q. gregaria*.

Malvin subprovince. This subprovince occupies the southernmost part of the continental shelf of the southwest Atlantic. The southwestern boundary of the subprovince is poorly defined and lies adjacent to the South Patagonian subprovince on the inner shelf at about 52°S. On the outer shelf, because of the north flowing Malvin Current, the fauna and consequently the subprovince is swept northward. Its western boundary lies along the 80-100m isobath. The subprovince embraces the area around the Islas Malvinas and covers the outer shelf and upper slope. The variety of bathymetric zones covered by this subprovince results in a mixture of species which might be considered unusual under conditions of more uniform depth.

The Malvin subprovince is devoid of any Brazilian elements. However, it contains many species in common with the Patagonian subprovinces, although the representatives of these species are almost always smaller and less well developed in the Patagonian subprovinces. There are some species which are particularly characteristic of the subprovince: *Angulogerina angulosa angulosa, Buccella peruviana,* f. campsi (large specimens), *Buliminella seminuda, Cassidulina crassa,* f. typica, *Cassidulinoides parkerianus, Discorbis isabelleanus, Ehrenbergina pupa, Hero-*

6

nallenia kempii, *Pullenia subcarinata subcarinata* and *Uvigerina bifurcata*.

The faunal assemblage near Cape Horn is somewhat distinct from that of the area around the Islas de los Estados and the Islas Malvinas. The Cape Horn area fauna has many *Anomalina vermiculata*, *Discorbis isabelleanus*, *D. williamsoni*, s.l., *Heronallenia kempii* and *Pullenia subcarinata subcarinata*, whereas the assemblage lying further to the northeast around the Islas de los Estados and the Islas Malvinas possesses more *Angulogerina angulosa angulosa*, *Cassidulina crassa*, *Cassidulinoides parkerianus*, *Ehrenbergina pupa* and *Uvigerina bifurcata*. These differences have been cited as sufficient to create two distinct subprovinces (Herb, 1971). However, they are minor compared to the strong degree of commonality between the two areas and we prefer to maintain them both as part of the Malvin subprovince.

Because the geographic limits of the Malvin subprovince are controlled in large part by the distribution of the Malvin Current on the outer shelf and because the limits of the Patagonian subprovinces seem to be influenced by the coastal water masses over the inner shelf, we have shown the distribution of species typical of these zones on separate sets of maps (Figs. 4-12 for the coastal waters and Figs. 13-17 for the Malvin Current).

SUMMARY OF BENTHIC FORAMINIFERAL INVESTIGATIONS IN THE AREA

The initial study of foraminifera from the southwest Atlantic shelf was published in 1839 by d'Orbigny as part of his classic work "Voyage dans l'Amérique Méridionale". Among the samples analyzed by d'Orbigny were littoral sands from the Patagonian coast between Bahía San Blas and the Golfo San Jorge, bottom samples from the nearshore waters around the Islas Malvinas, and a sample taken "en vue de terre au Cap Horn" at a depth of 160m. D'Orbigny described 52 species from the two sites mentioned above on the Argentine inner shelf (19 from the Patagonian coast, 38 from the Islas Malvines, with 5 common to both sites). From Cape Horn he described 5 species, 4 of which have proven to be common on the Argentine shelf and the fifth to be common on the western coast of South America.

After d'Orbigny's classic works of 1839, the next monumental study of Recent foraminifera was that published by Brady (1884) as a result of his study of material from the *Challenger* Expedition. Although almost a century has passed since the publication of this work, it has lost none of its importance, even despite the many taxonomic changes in the names proposed by Brady. Unfortunately, few *Challenger* sites occur on the southwest Atlantic shelf and those that do are rather poor from a faunal point of view. Only four stations are cited by Brady:

No. 313 off the Straits of Magellan at 100m, with 12 benthic foraminifera species (Murray, 1895).
No. 315 at Puerto Stanley, Islas Malvinas at 11m with "starved varieties of *Rotalia*, *Polystomella*, *Lagena* and *Bulimina*. The only species of particular interest were *Patellina corrugata* and *Bulimina elegantissima*" (p. 106).
No. 321 at the mouth of the Río de la Plata at 24m with ". . . various Miliolae, *Polystomella striatopunctata*, *Nonionina turgida* and *Haplophragmoides canariense*." (p. 167).
No. 322 at 60 miles east-south-east of Isla Lobos at 38m with no foraminifera.

Samples were also taken in the southwest Atlantic by the German research vessel *Gazelle*. Five of these samples, taken between 46-115m, were studied by Egger (1893) and yielded 9 species.

The Scottish National Antarctic Expedition to the Weddell Sea produced two stations on the Argentine shelf, one at Puerto Stanley, Islas Malvinas and another on the Burwood Bank. In the first of these, Pearcy (1914) found 6 species and in the second almost 80 species.

Cushman & Parker (1931) described more than 30 species from 8 samples from the Argentine shelf: 6 samples from the Islas Malvinas and one each from San Julián and Puerto Deseado.

One year later the classic work on the *Discovery* Expedition material appeared (Heron-Allen & Earland, 1932). All of the many samples examined by these investigators were taken on the southern part of the shelf south of 46°28'S. In all, 419 species were listed in this work, more than 10% of which were new. The character of the benthic foraminiferal fauna in the area was summarized as follows: "There is a monotonous sameness in the foraminiferal fauna over the whole of the shelf area, and on casual inspection it appears to be almost identical at the majority of stations and to consist of a few species only: *Cassidulina crassa*, *Cassidulina subglobosa*, *Cassidulina parkeriana*, *Ehrenbergina pupa*, *Uvigerina angulosa*, *Pullenia subcarinata*, *Truncatulina lobata*, *Truncatulina refulgens*, *Truncatulina ungeriana*, *Anomalina vermiculata*, *Pulvinulina karsteni*." (Heron-Allen & Earland, 1932, p. 295).

Two new species of *Silicotextulina* were established by Frenguelli (1935, 1947) from studies of material from San Blas and Quequén.

It was not until 1950 that a permanent site was established for the study of South American foraminifera. With the creation of the Laboratorio de Foraminíferos in the Museo Argentino de Ciencias Naturales "Bernardino Rivadavia" (Buenos Aires), the methodical study of the benthic foraminifera of the southwest Atlantic shelf began in earnest. About a decade later the Escola Geologia of the Universidad de Rio Grande do Sul in Puerto Alegre, Brazil established the Laboratorio Foraminiferologico. At this time the systematic study of southern Brazilian foraminifera began. The southern part of this area is included in this study.

Somewhat more than 50 works have been published since 1950 dealing with shelf foraminifera in the area. They are listed below by general category of investigation:

a. Foraminiferal systematics and distribution in euhaline

waters: Boltovskoy (1954b, c, 1962, 1963a), Boltovskoy & Lena (1966, 1970), Herb (1971) Lena (1966, 1974), Theyer (1966), Thompson (1978).

b. Zoogeography: Boltovskoy (1958a, 1964, 1970a, 1976), Herb (1971).

c. Ecology: Boltovskoy (1954a, 1956a, 1961, 1963b, c, 1966, 1971), Closs & Medeiros (1965), Scarabino (1967), Madeira-Falcetta (1974).

d. Oceanography: Boltovskoy (1959a, b, 1967, 1969, 1973), Boltovskoy & Lena (1969b), Giussani & Watanabe (in press), Lena (1976).

e. Microdistribution: Boltovskoy & Lena (1969c).

f. Cytology: Lena (1972), Lena & Freire (1974).

g. Foraminifera from mixohaline, brackish and fresh water: Boltovskoy (1957b, 1958a), Closs (1962), Closs & Barberena (1962a, b), Closs & Madeira (1962), Closs & Medeiros (1967), Wright (1968), Boltovskoy & Boltovskoy (1968), Boltovskoy & Lena (1971, 1974), Lena & L'hoste (1975), Boltovskoy & Giussani (in press).

h. Biological problems among foraminifera in various types of water: Boltovskoy (1956b, 1957a, 1965c), Forti & Roettger (1967), Closs & Madeira (1968), Boltovskoy & Lena (1969c).

i. Manuals and methodology: Boltovskoy (1953, 1958b, 1965a, b, d), Boltovskoy & Wright (1976).

These categories are somewhat arbitrary and several of the works cross the boundaries between them.

PRINCIPAL FACTORS INFLUENCING THE BENTHIC FORAMINIFERAL DISTRIBUTION

The specific ecology of benthic foraminifera is treated in Boltovskoy & Wright (1976) and data on environmental parameters from the area can be found in the various publications cited in the bibliography. Consequently we are going to discuss here only some general aspects of ecology and environmental relationships of the foraminifera with special reference to conditions in the southwest Atlantic shelf.

Benthic foraminifera seem to be most significantly affected by salinity, temperature, "depth", and total character of the water mass. Other factors may have considerable influence, particularly under local conditions, but are less well studied and understood. These factors include substrate character, nutrition, light intensity, turbidity, oxygen content, trace element distribution, etc.

Many of the factors listed above are dependent on each other and act in concert in their influence on foraminifera. It is often impossible to distinguish which single factor is the decisive one in most cases. For the sake of simplicity, we will treat the four major factors separately.

SALINITY

There are various ways of categorizing the salinity of natural waters. From the point of view of the observed distribution of benthic foraminifera the scheme given on Table 3 seems to work quite well.

When considering the distribution of foraminifera from the southwest Atlantic shelf these categories work well in delimiting assemblages. It is helpful to add the salt marsh to the categories given in Table 3. Although it is defined on the basis of dense vegetation and shallow waters rather than salinity, it and its fauna are strongly conditioned by the salinity which may climb as high as

Table 3. Classification of natural waters by salinity.

Salinity, $^0/_{00}$	Water type
40-75	hyperhaline
30-40	euhaline
18-30	hypohaline-mixohaline
0.5-18	hypohaline-brackish
<0.5	fresh

35‰ from its normal value less than 20‰. This variability is due to fluctuations between evaporation and precipitation.

The changes that occur from one foraminiferal assemblage to another as a result of salinity are gradual, and do not occur abruptly at the boundaries shown on Table 3. Nevertheless the boundaries shown on the table are useful ones for foraminifera. Particular groups of species tend to characterize each of these water types, although a given species may prosper in a water type of which it is not characteristic.

The diversity patterns of foraminifera vis-a-vis salinity are similar to those shown by other marine organisms, i.e., as the salinity decreases (or increases toward hyperhaline conditions) from euhaline conditions, the number of taxa decreases, but the number of specimens for each taxon increases.

On the southwest Atlantic shelf area we find benthic foraminiferal faunas characteristic of euhaline, hypohaline and fresh waters. There are no examples, of which we are aware, of hyperhaline faunas, but there are salt marsh assemblages in the northernmost part of the area.

Euhaline foraminifera. Those species now living on the continental shelf belong to this group. Their distribution is dependent more on temperature than on any other environmental factor. Of secondary importance are depth related parameters. The following publications deal with foraminifera of euhaline waters in this area: Boltovskoy (1954a-c, 1955, 1956a, b, 1957a, 1958b, 1959a, b, 1961, 1962, 1963a-c, 1964, 1965c, 1966, 1970a, b, 1971, 1976), Boltovskoy & Lena (1966, 1969a-c, 1970), Brady (1884), Cushman & Parker (1931), Egger (1893), Frenguelli (1935, 1947) Giussani & Watanabe (in press), Herb (1971), Heron-Allen & Earland (1932), Lena (1966, 1972, 1974), Lena & Freire (1974), d'Orbigny (1839), Pearcey (1914), Thompson (1978).

Hypohaline foraminifera. Foraminifera belonging to hypohaline assemblages occur in estuaries and lagoons in the area. They have been rather extensively studied and described in the following areas: Río Quequén (Boltovskoy & Boltovskoy, 1968; Wright, 1968), Río de la Plata (Boltovskoy, 1957b, 1958a; Boltovskoy & Lena, 1974),

Río Santa Lucía (Scarabino, 1967); Mar Chiquita lagoon (Lena & L'hoste, 1975), Lagoa dos Patos (Closs, 1962; Closs & Barbarena, 1962a, b; Closs & Madeira, 1968, Closs & Medeiros, 1965; Forti & Roettger, 1967; Madeira-Falcetta, 1974), Lagoa Mirim (Closs & Medeiros, 1967) Arroio Chuí (Closs & Barberena, 1962a; Closs & Madeira, 1962).

Fresh water foraminifera. To date, there are only four known localities in which living multilocular calcareous foraminifera have been found in fresh water ($<0.5\%_{oo}$) environments. These localities are the Río Paraná and Río de la Plata (Boltovskoy, 1958a; Boltovskoy & Lena, 1971; Boltovskoy & Giussani, in press), in Río Quequén (Boltovskoy & Boltovskoy, 1968) and in Lagoa dos Patos and Lagoa Mirim (Closs & Medeiros, 1965, 1967; Madeira-Falcetta, 1974).

Salt marsh foraminifera. Foraminifera belonging to this assemblage were described from a small area lying south of the Lagoa dos Patos (Madeira-Falcetta, 1974). In addition, there is an area of salt marsh lying on the southern edge of the estuary of the Río de la Plata along the margins of Bahía Samborombón. The foraminifera of this area were examined in 1968 by F. B. Phleger, whose work on salt marsh assemblages is well known. He concluded that the foraminifera of this area were not typical of salt marsh habitats and published nothing about them (personal communication, 1978).

TEMPERATURE

In shallow waters of normal marine salinity (euhaline), temperature plays the most significant role in determing the distribution of benthic foraminifera. On a generic level these thermal controls are well known. Tropic and sub-tropical areas are characterized by genera such as *Amphistegina*, *Archaias*, *Peneroplis*, *Heterostegina*, and *Borelis* whereas temperate waters exhibit such genera as *Buccella*, *Buliminella*, and *Cassidulina*. In high latitudes, these same shallow water environments show a marked increase in the proportion of agglutinated forms such as *Reophax*, *Trochammina* and *Alveolophragmium*. At the specific level the differences are even more marked. Within the genus *Elphidium*, for example, the species *E. sagrum* and *E. discoidale* are typical of warm water whereas the genus is represented in higher latitudes by *E. macellum*, *E. bart-*

letti, *E. crispum*, and *E. arcticum*. Not all species are so significantly affected by temperature changes and some, like *Quinqueloculina seminulum* are cosmopolitan.

The area studied in this atlas lies generally along longitudinal lines and consequently there is a gradual increase in water temperature on the inner shelf from south to north. Seasonal values along the shelf are shown in Table 2.

DEPTH

Depth per se is not a factor which controls the distribution of benthic foraminifera, although depth related factors may have a significant influence. The relationships between these depth related factors is not clearly understood and it is often easier and more convenient to outline the distribution of various taxa as a function of depth than it is to unravel the several interrelated factors that are actually exhibiting the control. That there are differences in faunal assemblages with depth is not disputed. The shelf species of such genera as *Elphidium*, *Ammonia*, *Quinqueloculina* and *Bolivina* give way on the slope to species of such genera as *Uvigerina*, *Cassidulina*, *Pleurostomella* and *Gyroidina*.

WATER MASS

In the shelf areas of the southwest Atlantic, the distribution and characteristics of water masses play an extremely important role in the distribution of benthic foraminifera.

For some time now, it has been observed that plankton distribution is strongly correlated with masses of water. The first suggestions that benthic foraminifera could serve as indicators of water mass identity and movement occurred during the early part of this century. However, these suggestions were made in passing and were not the result of detailed studies devoted to the subject. The first specific study of benthic foraminifera as hydrologic indicators was conducted by Boltovskoy (1959a, b) who utilized them to delimit the western deep boundary of the Malvin Current in the northern part of the Argentine continental shelf. The particular distribution of the benthic foraminifera of the inner shelf of the southwest Atlantic also testifies to the influences of water masses on the fauna.

SYSTEMATICS AND DISTRIBUTION

The classification used in this work is essentially that of Loeblich & Tappan (1964, 1974) as it is the best known and most easily used. Nevertheless we often find ourselves in favor of a generic concept which is somewhat more conservative than that proposed by Loeblich and Tappan. We think that the proliferation of taxa (even at the generic level) as reflected in Loeblich & Tappan (op. cit.) and followed by many other workers, is prejudicial to the practical interests of the science (Boltovskoy, 1965d). Consequently, we will not always rigidly follow the Loeblich and Tappan classification, but will sometimes use other generic names, either to maintain a somewhat broader interpretation, or to follow the principle of conservation rather than that of priority (Todd, 1963; Le Calvez, 1969; Boltovskoy & Wright, 1976). Insofar as species identification is concerned we have expended a great deal of energy, not only at the microscope, but also in searching the literature and examining original type material, to be as precise and faithful to the original descriptions as possible.

When reading the descriptions that follow, the reader should consider several points.

1. All species descriptions given here are based only on specimens from the southwest Atlantic shelf area.
2. The informal category of forma is used here because it provides a convenient manner of expressing differences within a population without having to erect a formal taxon. Because this category lies outside the International Rules of Zoological Nomenclature, it is written in roman rather than italic type, and separated by a comma from the formal name.
3. Generic traits are given separate from those of species in order to save space and avoid redundancy in species descriptions.
4. All the species described in this atlas are classified according to their preference for euhaline, hypohaline, or fresh water habitats. The geographic distribution of euhaline species is shown in Figs. 4-17. The solid lines representing the distribution of these species are schematic and generalized. They do not indicate that each species has

been found at all points included within the extreme limits of the line, but only that the limits of the line mark the general boundaries of environmental conditions that are favorable for the species in question. The absence of more precise data is due to the scarcity of samples or limited abundance of some species. The northern and southern limits of the lines on the figures coincide with the points beyond which well developed and preserved specimens were not found. Occurrences of pathologic or taxonomically suspect specimens were not considered in the figure preparation. Quantitative abundances of species are not shown on these figures.

The distribution of hypohaline and fresh water forms is discussed separately for each such species in the text.
5. The locations of all sites mentioned in the text are shown on Fig. 3.
6. The euhaline species are described separately from those occurring only in hypohaline and fresh waters. Within each group the species are listed alphabetically. The systematic position of the genera is given in Table 4.
7. The species in the area were divided into three broad categories based on frequency of occurrence. For those 100 species whose frequency of occurrence could be characterized as abundant, common or frequent a random sample of at least 30 specimens was selected and measured to determine range of sizes, mean size and the coefficient of variation (standard deviation/mean). For those species which were less than frequent yet neither rare nor isolated, a representative suite of specimens was measured and the range of sizes calculated. There were 91 such species. The remaining species occur only in limited numbers or in isolated sites in the area. These species are neither described nor listed here.

All measurements are given in μm.
8. Finally we need to point out that the geographic distributions and environmental limits given for any species are merely our best estimate at the moment. The area covered in this study is quite large and despite the years of effort in collecting samples there are still areas that are poorly studied. We hope that future investigations in this area will fill these gaps.

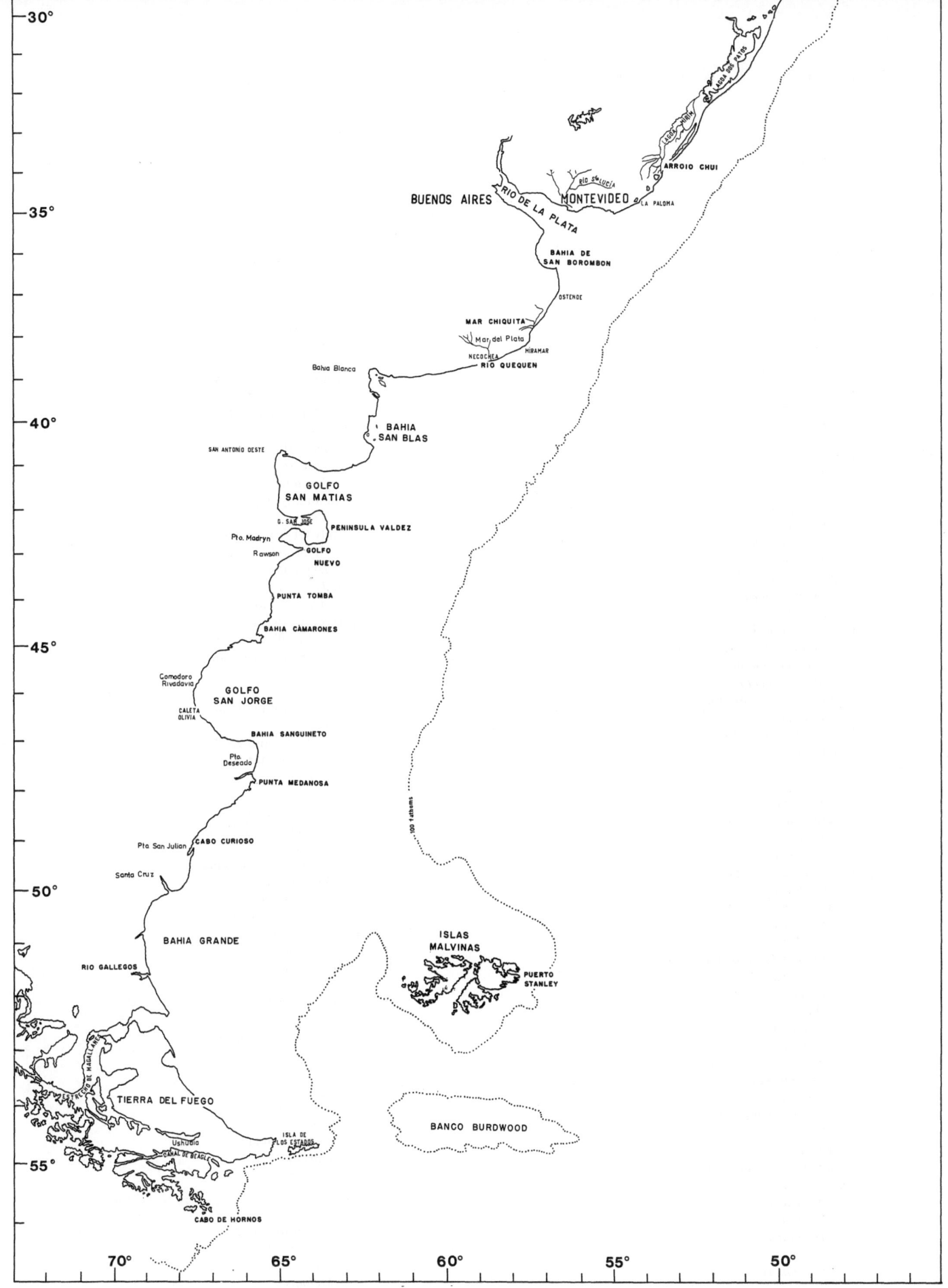

Fig. 3.   Geographic localities cited in the text

13

Table 4. Systematic position of the most common genera found in the area.

Phylum Protozoa
  Class Sarcodina
    Subclass Rhizopoda
      Order Foraminiferida
        Suborder Allogromiina
          Family Allogromiidae
            Allogromia
        Suborder Textulariina
          Superfamily Ammodiscacea
            Family Saccamminidae
              Subfamily Psammosphaerinae
                Psammosphaera
              Subfamily Saccammininae
                Dahlgrenia
                Saccammina
          Superfamily Lituolacea
            Family Hormosinidae
              Protoschista
              Reophax
            Family Rzehakinidae
              Miliammina
            Family Lituolidae
              Subfamily Haplophragmoidinae
                Cribrostomoides
                Haplophragmoides
                Recurvoides
              Subfamily Lituolinae
                Ammoscalaria
                Ammotium
            Family Textulariidae
              Subfamily Spiroplectammininae
                Spiroplectammina
                Morulaeplecta
              Subfamily Textulariinae
                Textularia
            Family Trochamminidae
              Subfamily Trochammininae
                Arenoparrella
                Jadammina
                Trochammina
              Subfamily Remaneicinae
                Remaneica
        Suborder Miliolina
          Family Fischerinidae
              Cyclogyra
          Family Nubeculariidae
              Spiroloculina
          Family Miliolidae
            Subfamily Quinqueloculininae
              Massilina
              Pyrgo
              Quinqueloculina
              Sigmoilina
              Triloculina
            Subfamily Miliolinellinae
              Biloculinella
              Miliolinella
            Subfamily Tubinellinae
              Tubinella
        Suborder Rotaliina
          Superfamily Nodosariacea
            Family Nodosariidae
              Amphicoryna
              Astacolus
              Dentalina
              Lagena
              Orthomorphina
              Robulus
            Family Polymorphinidae
              Globulina
              Guttulina
              Sigmomorphina

            Family Glandulinidae
              Fissurina
              Oolina
          Superfamily Buliminacea
            Family Turrilinidae
              Buliminella
            Family Bolivinitidae
              Bolivina
            Family Islandiellidae
              Cassidulinoides
            Family Buliminidae
              Bulimina
              Virgulina
            Family Uvigerinidae
              Angulogerina
              Hopkinsina
              Uvigerina
          Superfamily Discorbacea
            Family Discorbidae
              Subfamily Discorbinae
                Buccella
                Discorbinella
                Discorbis
                Epistominella
              Subfamily Baggininae
                Cancris
            Family Glabratellidae
              Glabratella
              Heronallenia
            Family Asterigerinidae
              Asterigerinata
          Superfamily Spirillinacea
            Family Spirillinidae
              Subfamily Spirillininae
                Mychostomina
                Spirillina
              Subfamily Patellininae
                Patellina
          Superfamily Rotaliacea
            Family Rotaliidae
              Ammonia
              Rolshausenia
            Family Elphidiidae
              Subfamily Elphidiinae
                Eliphidium
              Subfamily Faujasininae
                Cribrorotalia
                Notorotalia
          Superfamily Orbitoidacea
            Family Eponidae
              Poroeponides
            Family Cibicididae
              Cibicides
            Family Planorbulinidae
              Planorbulina
          Superfamily Cassidulinacea
            Family Loxostomidae
              Loxostomum
            Family Cassidulinidae
              Cassidulina
              Ehrenbergina
            Family Nonionidae
              Florilus
              Nonion
              Nonionella
              Pullenia
            Family Anomalinidae
              Anomalina
              Hanzawaia
              Melonis
          Superfamily Robertinacea
            Family Ceratobuliminidae
              Hoeglundina

EUHALINE SPECIES

ALLOGROMIA Rhumbler, 1904

Test free, unilocular, ovate to spherical; wall protein-aceous, delicate and sometimes agglutinated, light yellow to green-brown or colorless; aperture round, simple, terminal, with short entosolenian tube.

Allogromia flexibilis (Wiesner), Pl. 1, figs. 1-3.
ORIGINAL CITATION: *Technitella flexibilis* Wiesner, 1931. In: Drygalski, Deutsch Südpol.-Exp., v. 20 (Zool., v. 12), p. 85, pl. 7, fig. 75.
TYPE LOCALITY: Gauss Station: 66°02′S, 89°38′E.
AGE: Recent.
DESCRIPTION: Test circular in transverse section, flexible when wet, collapsed when dry; wall finely agglutinated with organic cement; aperture small, atop a short neck.
LENGTH: range: 320-980.
DISTRIBUTION: see Figure 4.

AMMONIA Brünnich, 1772

Test free, biconvex, coiled in low trochospiral of 3-4 volutions; sutures thick, depressed, radial on umbilical side, curved slightly backward on spiral side; wall calcareous, radial, double walled along septa, finely perforate with irregular pustules of calcite located along the sutures of the umbilical side as well as on the umbonal area; internal margins of sutures break up into fissures and may form a central knob in the umbilicus; aperture interiomarginal.

Ammonia beccarii (Linné), Pl. 1, figs. 4-7.
ORIGINAL CITATION: *Nautilus beccarii* Linné, 1758. *Systema Naturae*, ed. 10, Holmiae, v. 1, p. 710, figured by Plancus, Conchiolog. pl. 1, figs. 1a-c.
TYPE LOCALITY: Mediterranean Sea, probably at Rimini, Italy.
AGE: Recent (Pliocene, if from Rimini).
DESCRIPTION: Test circular, periphery rounded; chambers slightly inflated, ending in umbilical extension, 9-12 in final whorl; wall light yellow or white, highly ornamented, shiny, finely perforate, with large tubercules (Pl. 1, fig. 7) located along sutures and becoming reduced in last whorl of spiral side; sutures flush and limbate in early portion of spiral side, depressed and open in last whorl on both sides; aperture a narrow slit at base of apertural face opening into umbilicus.
DIAMETER: range: 410-1420; mean: 800; coefficient of variation: 0.29.
DISTRIBUTION: see Figure 4.

Ammonia ex. gr. *A. parkinsoniana* (d'Orbigny), Pl. 1, figs. 8-9.

ORIGINAL CITATION: *Rosalina parkinsoniana* d'Orbigny, 1839, In: de la Sagra, Hist. Phys. Polit. Natur. Cuba, p. 99, pl. 4, figs. 25-27.
TYPE LOCALITY: European and Cuban coasts.
AGE: Recent.
DESCRIPTION: Test outline circular, peripheral margin rounded; $2\frac{1}{2}$ volutions, the last of which contains 8 chambers; wall yellow to yellow-brown, coarsely perforate; on the spiral side, septal sutures limbate, spiral sutures depressed in the final whorl; on umbilical side, sutures depressed and leading into an umbilical cavity which may be filled with a knob; aperture a narrow slit at base of last chamber.
DIAMETER: range: 210-450; mean: 310; coefficient of variation: 0.19.
OBSERVATIONS: *Ammonia* ex gr. *A. parkinsoniana* is undoubtedly very closely related to *A. beccarii*. Nevertheless we have not observed any forms which are transitional between the two and we are inclined to keep them as separate taxa. In relation to *A. beccarii*, this species has a smaller test, larger and more distinct pores, a distinct tendency to form a central cone on the spiral side, a more lobed periphery, more globose chambers, and less well developed ornamentation. The geographic distribution of the two species overlaps but *A.* ex gr. *A. parkinsoniana* appears to prefer somewhat more brackish conditions although it thrives in water of normal salinity also. *A. parkinsoniana* was redescribed by Le Calvez (1977) from the original material of d'Orbigny. Our specimens are quite similar and yet have a much greater porosity and for this reason we are somewhat doubtful of its designation.
DISTRIBUTION: see Figure 4. This species also occurs in waters with salinities less than normal (Lagoa dos Patos, Lagoa Mirim, Arroio Chuí, Río de la Plata and Mar Chiquita).

AMPHICORYNA Schlumberger, 1881

Test free, elongate, uniserial, the early chambers compressed and partially coiled, adult chambers rectilinear; wall calcareous, radial and lamellar, perforate, costate or smooth; aperture circular, terminal, central, radiate, on top of a long neck.

Amphicoryna scalaris (Batsch), Pl. 1, figs. 10-12.
ORIGINAL CITATION: *Nautilus (Orthoceras) scalaris* Batsch, 1791. Conchyl. Seesand., p. 1, pl. 2, fig. 4.
TYPE LOCALITY: Italian coast.
AGE: Not given.
DESCRIPTION: Test with rounded base, 2-3 globose chambers; wall translucent, lustrous and finely perforate, smooth or with striae; sutures depressed and easily seen; aperture circular and irregularly radiate, located at end of costate neck, with a lip.

15

LENGTH: range; 210-640; mean: 360; coefficient of variation: 0.36.

DISTRIBUTION: see Figure 4.

ANGULOGERINA Cushman, 1927

Test free, elongate, triangular in transverse section, triserial with tendency toward uniserial; wall calcareous, finely perforate, radial, with longitudinal costae; aperture terminal on a neck with lip.

Angulogerina angulosa angulosa (Williamson), Pl. 1, figs. 13-16.

ORIGINAL CITATION: *Uvigerina angulosa* Williamson, 1858. Rec. Foram. Gr. Brit., Ray Soc., p. 67, pl. 5, fig. 140.

TYPE LOCALITY: British Isles.

AGE: Recent.

DESCRIPTION: Test ovate-elongate; chambers irregular, inflated, covered with well developed costae which are not continuous across sutures, the central angular costae gives each chamber its triangular shape (Pl. 1, fig. 15); sutures depressed; wall translucent, white, finely perforate; aperture circular, on short neck, with tooth.

LENGTH: range: 310-770; mean: 520; coefficient of variation: 0.23.

OBSERVATIONS: This is the dominant species of the Malvin Current.

DISTRIBUTION: see Figures 4 and 13.

Angulogerina angulosa occidentalis (Cushman), Pl. 1, figs. 17-18.

ORIGINAL CITATION: *Uvigerina occidentalis* Cushman, 1923. U.S. Nat. Mus., Bull. 104, pt. 4, p. 169.

TYPE LOCALITY: Tortugas region, West Indies.

AGE: Recent.

DESCRIPTION: This subspecies differs from the *nominat subspecies* in that it is markedly angulate only near the apertural end, early chambers are rounded and covered with small rounded costae; final chambers are less well ornamented, and it is smaller than the *nominat subspecies*.

LENGTH: range: 200-430; mean: 300; coefficient of variation: 0.20.

OBSERVATIONS: We believe this subspecies represents a geographic isolate of *A. angulosa* adapted to shallower water with less salinity than is normal for the species.

DISTRIBUTION: see Figure 4.

ANOMALINA d'Orbigny, 1826

Test approximately planispiral, the internal spire of the dorsal side produces an umbo, ventral side depressed; wall calcareous, perforate; aperture interiomarginal extending slightly onto dorsal side of periphery.

Anomalina vermiculata (d'Orbigny), Pl. 1, figs. 19-21.

ORIGINAL CITATION: *Truncatulina vermiculata* d'Orbigny, 1839, Voy. Amér. Mérid., v. 5, pt. 5, p. 39, pl. 6, figs. 1-3.

TYPE LOCALITY: Islas Malvinas and Cabo de Hornos.

AGE: Recent.

DESCRIPTION: Test suborbicular, peripheral margin generally rounded with slight degree of angularity on ventral side, dorsal side with variable convexity, periphery lobate, final whorl with 7-8 chambers; partially evolute on umbilical side; wall thick, containing many large pores concentrated on both sides or along peripheral margin only; sutures depressed; aperture a slit with lip.

DIAMETER: range: 430-1260; mean: 850; coefficient of variation: 0.28.

OBSERVATIONS: Our specimens match very well with those of d'Orbigny housed in the Museum d'Histoire Naturelle, Paris.

DISTRIBUTION: see Figures 4 and 13.

ASTACOLUS Montfort 1808

Test free, elongate, arcuate, compressed; chambers numerous, low, wide and added along the axis of an open planospire; sutures oblique and elevated on external peripheral margin; aperture terminal, radiate, rounded.

Astacolus crepidulus (Fichtel and Moll), Pl. 2, figs. 1-2.

ORIGINAL CITATION: *Nautilus crepidulus* Fichtel and Moll, 1798. *Testac. Microsc.*, p. 107, pl. 19, figs., g-i.

TYPE LOCALITY: Livorno coast, Tuscany, Italy.

AGE: Recent.

DESCRIPTION: Test expanding, internal peripheral margin rounded, external margin subangular; chambers, 8-12, first few added in closed spire and later ones added in uniserial manner; sutures slightly depressed; aperture terminal and radiate.

LENGTH: range: 290-890.

DISTRIBUTION: see Figure 4.

ASTERIGERINATA Bermúdez, 1949

Test free, trochospiral, compressed, with 3 volutions; spiral side convex and umbilical side planar; secondary chambers form a small globose rosette about the umbilical plug; wall glassy; aperture short, elliptical, located along margin of last chamber of umbilical side.

Asterigerinata pacifica Uchio, Pl. 2, figs. 3-5.

ORIGINAL CITATION: *Asterigerinata pacifica* Uchio, 1960. Cushman Found, Foram. Res., Sp. Publ. 5, p. 67, pl. 10, figs. 26-31.

TYPE LOCALITY: San Diego, California.

AGE: Recent.

DESCRIPTION: Test small, circular with rounded peripheral

margin, last whorl with 6-7 chambers; secondary chambers on umbilical side form a central rosette at the external extremities of which are small orifices; sutures on spiral side unclear, on umbilical side slightly depressed; wall thin, translucent; aperture large, irregular, ovate and occupying large part of last chamber.

DIAMETER: range; 150-850.

DISTRIBUTION: see Figure 4.

BILOCULINELLA Wiesner, 1931

Test free, ovate; microspheric form with first few chambers quinqueloculine, later triloculine and finally biloculine; megalospheric forms often biloculine throughout their ontogeny; wall calcareous, imperforate, porcellaneous; aperture terminal with valvular tooth.

Biloculinella irregularis (d'Orbigny), Pl. 2, figs. 6-8.

ORIGINAL CITATION: *Biloculina irregularis* d'Orbigny, 1839. Voy. Amér. Mérid., v. 5, pt. 5, p. 67, pl. 8, figs. 20, 21.

TYPE LOCALITY: Islas Malvinas.

AGE: Recent.

DESCRIPTION: Test height and thickness somewhat greater than width; peripheral margin rounded; chambers globular, compressed laterally; wall smooth, lustrous and white; sutures quite visible, slightly depressed; aperture large, semicircular with concave valvular flap.

LENGTH: range: 380-550.

OBSERVATIONS: The types for this species are missing from the d'Orbigny collection in Paris.

DISTRIBUTION: see Figures 4 and 13.

BOLIVINA d'Orbigny, 1839

Test free, elongate, generally compressed, biserial; wall calcareous, perforate, radial, smooth or ornamented, with marginal keel in some species; aperture elongate, curving up the apertural face from the suture to near the terminal position, one margin with a lip of variable thickness which continues within the chamber as a tooth plate.

Bolivina compacta Sidebottom, Pl. 2, figs. 9, 10, 12.

ORIGINAL CITATION: *Bolivina robusta* Brady, var. compacta Sidebottom, 1905. Manchester Lit. Phil. Soc., Mem. 48-53, p. 15, pl. 3, fig. 7.

TYPE LOCALITY: Island of Delos.

AGE: Recent.

DESCRIPTION: Test margins generally parallel with slight expansion of final chambers giving triangular outline to whole test; periphery rounded, apical end acuminate, apertural end bilobate, composed of 6-9 pairs of convex chambers; wall translucent to transparent, shiny, with numerous large pores; sutures straight, depressed, making a 65° angle with growth axis; aperture elongate and cov-

ering major portion of apertural face.

LENGTH: range: 160-470; mean: 270; coefficient of variation: 0.30.

DISTRIBUTION: see Figure 4. This species is also recorded in brackish waters of the Río de la Plata and Río Quequén.

Bolivina cf. *B. danvillensis* Howe and Wallace, Pl. 2, figs. 11, 13-15.

ORIGINAL CITATION: *Bolivina danvillensis* Howe and Wallace, 1932. Louisiana Dept. Cons. Geol., Bull. 2, p. 56, pl. 11, figs 8a, b.

TYPE LOCALITY: Danville Landing, Louisiana.

AGE: Late Eocene.

DESCRIPTION: Test elongate, somewhat flattened, with nearly parallel lateral margins; peripheral margin rounded, occasionally acute and lobate; chambers flattened in early portions of test becoming more inflated in adult portion, 9-10 pairs; wall shiny with many small pores on lower half of each chamber (Pl. 2, fig. 14); sutures oblique, very slightly curved and making 90° angle with growth axis, depressed, with weak reentrant; aperture elongate with thick raised rim along one of its borders.

LENGTH: range: 170-380.

OBSERVATIONS: Our specimens differ from those described by Howe and Wallace in that they are more flattened laterally, have a more rounded periphery, and less lobate sutures.

DISTRIBUTION: see Figure 4.

Bolivina difformis (Williamson), Pl. 2, figs. 16, 17.

ORIGINAL CITATION: *Textularia variabilis*, var. difformis, Williamson, 1858. Rec. Foram. Gr. Brit., Ray Soc., p. 77, pl. 6, figs. 166, 167.

TYPE LOCALITY: British Isles.

AGE: Recent.

DESCRIPTION: Test small, compressed, maximum length corresponds to medial axis, lanceolate; periphery serrate, acute, particularly in adult portion where chambers end in denticles; chambers rounded in early portion, without denticles, 8-10 pairs in adult form; wall shiny, finely perforate; sutures depressed and arcuate; aperture ovate.

LENGTH: range: 230-270.

DISTRIBUTION: see Figure 4.

Bolivina doniezi Cushman and Wickenden, Pl. 2, figs. 18-20.

ORIGINAL CITATION: *Bolivina doniezi* Cushman and Wickenden, 1929. U.S. Nat. Mus., Proc., v. 75, art. 9, p. 9, pl. 4, figs. 3a, b.

TYPE LOCALITY: Cumberland Bay, Juan Fernandez Island.

AGE: Recent.

DESCRIPTION: Test small, peripheral margin rounded; chambers in early part of test small and narrow, later becoming more inflated, 7-9 pairs; wall shiny with pores which are more distinct in apertural region and/or on lower half of each chamber; sutures depressed and curved; aperture oval, elongate, with flanged margin, of intermediate size for the genus.

LENGTH: range: 140-270; mean: 200; coefficient of variation: 0.20.

DISTRIBUTION: see Figure 4. This species also occurs in the brackish waters of the Río de la Plata.

Bolivina ordinaria Phleger and Parker, Pl. 3, figs. 1-3.

ORIGINAL CITATION: *Bolivina simplex* Phleger and Parker, 1950, Geol. Soc. Amer. Mem. 46, p. 14, pl. 7, figs. 4, 5a, b, 6 (emend. Phleger and Parker, 1952. Curshman Found. Foram. Res., Contr., v. 3, pt. 1, p. 14, *Bolivina ordinaria*).

TYPE LOCALITY: Northwest Gulf of Mexico.

AGE: Recent.

DESCRIPTION: Test compressed, acute angle at apertural end, rounded at apical end, maximum length and thickness along medial line, elliptical in transverse section; chambers narrow and elongate, 7-10 pairs; wall translucent to opaque, shiny, appears to be thicker along sutures; sutures depressed, wide, curved, making 50-60° angle with growth axis; aperture narrow, small, extending from basal suture to terminal position.

LENGTH: range: 160-340; mean: 260; coefficient of variation: 0.19.

DISTRIBUTION: see Figures 4 and 13.

Bolivina pseudoplicata Heron-Allen and Earland, Pl. 3, figs. 4-8.

ORIGINAL CITATION: *Bolivina pseudoplicata* Heron-Allen and Earland, 1930. Jour. Roy. Micr. Soc., v. 50, p. 81, pl. 3, figs. 36-40.

TYPE LOCALITY: Plymouth area, England.

AGE: Recent.

DESCRIPTION: Test broad at apertural end, peripheral margin subacute to rounded, chambers somewhat inflated, 7-8 pairs; wall opaque, white, sometimes translucent, with a reticulate pattern of randomly distributed pores bounded by raised polygonal borders (Pl. 3, fig. 8), a pattern which ceases on upper part of final two chambers; final two chambers covered with elongate, discontinuous rugae alligned parallel to border of aperture, remainder of test marked by two prominent longitudinal crests which form a zig-zag pattern parallel to the long axis and outline a central furrow; crests radiate laterally and outline the depressed sutures which are somewhat obscure; aperture ovate.

LENGTH: range: 190-470; mean: 320; coefficient of variation: 0.22.

DISTRIBUTION: see Figure 4. In addition, this species is found in the brackish waters of the Río de la Plata and Rio Quequén.

Bolivina striatula Cushman, Pl. 3, figs. 9-13.

ORIGINAL CITATION: *Bolivina striatula* Cushman, 1922. Carnegie Inst., Publ. 311, p. 27 (pars), pl. 3, fig. 10.

TYPE LOCALITY: Dry Tortugas, Florida.

AGE: Recent.

DESCRIPTION: Test compressed, rather long, peripheral margin subangulate; apical end acute and may posses spine, oral end wide and rounded; 7-10 pairs of chambers; wall shiny, translucent, finely perforate at apical end covered with fine costae which reduce to striae as they run the length of test; sutures depressed, forming an angle of 60° with growth axis, slightly curved along the peripheral margin; aperture ovate, of medium size for genus.

LENGTH: range: 270-550; mean: 390, coefficient of variation: 0.21.

DISTRIBUTION: see Figure 4. This species is able to tolerate reductions in salinity and can be found in the brackish waters of Lagoa dos Patos, Arroio Chuí, and the area near the mouth of the Río de la Plata.

Bolivina tortuosa Brady, Pl. 3, figs. 14-17.

ORIGINAL CITATION: *Bolivina tortuosa* Brady, 1881. Quart. Jour. Micr. Sci., v. 21, p. 51, 1884, Challenger Exp., Repts., Zool., v. 9, p. 420, pl. 52, figs. 31-34.

TYPE LOCALITY: Not designated (probably in the South Pacific).

AGE: Recent.

DESCRIPTION: Test torted, apical end narrow and rounded, apertural end wide, peripheral margin with well developed thick, somewhat lobate keel near apertural end; chambers flattened, only slightly inflated, growing in loose spiral about major axis, 7-8 pairs; wall thick, translucent, shiny, with large pores at base of a series of pits which give the test a rugose aspect; sutures depressed, rather wide; aperture elongate, narrow.

LENGTH: range: 210-320.

DISTRIBUTION: see Figure 4.

Bolivina translucens Phleger and Parker, Pl. 3, figs. 18-21.

ORIGINAL CITATION: *Bolivina translucens* Phleger and Parker, 1951. Geol. Soc. Amer., Mem. 46, p. 15, pl. 7, figs. 13a, b.

TYPE LOCALITY: Northwest Gulf of Mexico.

AGE: Recent.

DESCRIPTION: Test elongate, narrow, peripheral margin lobate and rounded; chambers slightly inflated and wider than high, 6-7 pairs, proloculus large; wall translucent, smooth, shiny, finely perforate on lower half of each chamber; aperture ovate, narrow and small.

LENGTH: range: 210-360.

DISTRIBUTION: see Figure 4.

Bolivina variabilis (Williamson), Pl. 4, figs. 1-4.

ORIGINAL CITATION: *Textularia variabilis* Williamson,

1858. Ray Soc., p. 75, pl. 6, figs. 162, 163.

TYPE LOCALITY: British Isles.

AGE: Recent.

DESCRIPTION: Test acute at apical end, rounded at apertural end, peripheral margin rounded; chambers somewhat inflated, 7-8 pairs; wall translucent with numerous widely distributed pores each surrounded by a polygonal rim (Pl. 4, fig. 4) which gives the test a rugose aspect; central zig-zag suture is wide and depressed below surface of test; aperture ovate with a thick rim.

LENGTH: range: 200-410; mean: 310; coefficient of variation: 0.16.

DISTRIBUTION: see Figure 4.

BUCCELLA Anderson, 1952

Test free, trochospiral, planate to biconvex, peripheral margin rounded to keeled; wall calcareous, perforate; sutures and umbilical zone covered with granules of varying development; principal aperture interiomarginal, lying between umbilicus and periphery; supplementary apertures along sutures of umbilical side.

Buccella peruviana (d'Orbigny), sensu lato, Pl. 4, figs. 5-22.

ORIGINAL CITATION: *Rotalina peruviana* d'Orbigny, 1839. Voy. Amér. Mérid., v. 5, pt. 5, p. 35, pl. 2, figs. 3-5.

TYPE LOCALITY: West coast of South America.

AGE: Recent.

DESCRIPTION: Test circular in plan view, spiral side convex, umbilical side planar to slightly convex, peripheral margin smooth to lobate, with or without a rim which can be wide or delicate; chambers, 6-11 in final whorl; wall smooth, shiny, opaque to translucent; sutures on spiral side obscure, tangential, curved, on umbilical side radial, markedly depressed or excavated; aperture obscure, generally covered with calcareous granules; supplementary apertures small.

OBSERVATIONS: Assemblages of this species are composed of three distinct end members which intergrade completely with one another.

Buccella peruviana (d'Orbigny), forma typica, Pl. 4, figs. 5, 6, 10, 11, 16, 17.

CHARACTERISTIC TRAITS: Large size; biconvex test; 11 chambers in the final whorl; umbilical sutures radial, excavated and covered with fine dense granules; wide circular umbilicus filled with granules; peripherial margin with strongly developed rim.

DIAMETER: range: 400-500.

OBSERVATIONS: This forma was originally described from Callao, Peru. It entered the Atlantic area via the subantarctic waters of the Malvin Current but in forms with fewer chambers in the final whorl (9-10). Good examples can be found in the original d'Orbigny material in Paris.

DISTRIBUTION: see Figure 13.

Buccella peruviana (d'Orbigny), forma campsi, Pl. 4, figs. 7, 8, 12, 13, 18, 19.

ORIGINAL CITATION: *Eponides peruvianus campsi* Boltovskoy, 1954, Inst. Nac. Inv. Cienc. Nat., Rev., Geol., v. 3, n. 3, p. 265, pl. 17, figs. 6-8.

CHARACTERISTIC TRAITS: This forma is distinguished from forma typica by its smaller size (but larger than forma frigida described below), fewer number of chambers in the final whorl (7-9), and a less well developed marginal flange.

DIAMETER: range: 280-630; mean: 400; coefficient of variation: 0.20.

OBSERVATIONS: This forma is typical of the external shelf in the zone of the Malvin Current.

DISTRIBUTION: see Figure 13. Small specimens of this forma are also found in brackish waters of Lagoa dos Patos and Arroio Chuí.

Buccella peruviana (d'Orbigny), forma frigida, Pl. 4, figs. 9, 14, 15, 20-22.

ORIGINAL CITATION: *Pulvinulina repanda* Fichtel and Moll, var. karsteni Reuss. Parker and Jones. Phil. Trans. Roy. Soc., v. 155, n. 1, p. 396, pl. 14, figs. 14, 15, 17; pl. 16, figs. 38-40. *Pulvinulina frigida* Cushman, 1921 (1922). Canad. Biol. Fish., Contr., p. 12.

CHARACTERISTIC TRAITS: This forma is characterized by having a very small test, 6-9 chambers in the final whorl, the absence of a peripheral rim, and narrow umbilical sutures. The granules are larger but less dense than in the other formae.

DIAMETER: range: 210-480; mean: 330; coefficient of variation: 0.24.

OBSERVATIONS: In the inner shelf regions from 32-52°S, forma frigida is the most abundant benthic foraminifer and is the only representative of *B. peruviana* in this area.

DISTRIBUTION: see Figure 5. This forma can tolerate brackish water and occurs in Lagoa dos Patos and Arroio Chuí.

BULIMINA d'Orbigny, 1826

Test free, elongate or pyriform, triserial; chambers inflated; wall calcareous, perforate, smooth or with longitudinal costae or spines; sutures of variable visibility; aperture ovate, extending from basal suture to terminal position, commonly lying in a depression, with a lip on one side, and a tooth plate.

Bulimina aculeata d'Orbigny, Pl. 5, figs. 1-3.

ORIGINAL CITATION: *Bulimina aculeata* d'Orbigny, 1826. Ann. Sci. Nat., sér. 1, v. 7, p. 269, no. 7; Fornasini, 1902, Acc. Sci. Bologna, ser. 5, v. 9, p. 153, fig. 4.

TYPE LOCALITY: Adriatic Sea near Rimini, Italy.

AGE: Probably Pliocene.

DESCRIPTION: Test acute at apical end and bearing one or two terminal spines rounded at apertural end; spire with 4-5 volutions; chambers inflated in last few whorls which occupy major part of test, with salient lower borders marked by short stubby spines; wall translucent, white, shiny, finely perforate; sutures depressed; aperture large.

LENGTH: range: 200-540; mean: 360; coefficient of variation: 0.19.

DISTRIBUTION: see Figure 5.

Bulimina elongata d'Orbigny, Pl. 5, figs. 4-6.

ORIGINAL CITATION: *Bulimina elongata* d'Orbigny, 1826. Ann. Sci. Nat., sér. 1, v. 7, p. 269; 1846. Foram Foss Bass. Tert. Vienne, p. 187, pl. 11, figs. 19, 20.

TYPE LOCALITY: Nussdorf, Vienna.

AGE: Tertiary, probably Miocene.

DESCRIPTION: Test ovate, uniform width throughout length; rounded at apertural end, lobate profile; spire with 6 (sometimes 5) volutions; chambers inflated, rounded, with 2½ chambers per whorl; wall translucent, shiny, smooth, finely perforate, early chambers may exhibit a slight rugose character; sutures depressed; aperture large.

LENGTH: range: 280-740; mean: 430; coefficient of variation: 0.21.

OBSERVATIONS: Although similar in many respects to *B. gibba*, this species has only 2½ chambers per whorl and consequently lacks the uniform rectilinear chamber disposition of *B. gibba*.

DISTRIBUTION: see Figure 5.

Bulimina gibba Fornasini, Pl. 5, figs. 7-9.

ORIGINAL CITATION: *Bulimina gibba* Fornasini, 1900. Mem. Real. Accad. Sci. Ist. Bologna, ser. 5, v. 8, p. 378, figs. 32, 34.

TYPE LOCALITY: Porto Corsini at the beach of Ravenna, Italy.

AGE: Recent.

DESCRIPTION: Test with marked expansion in the last whorl, narrower toward apex, lobate in profile; spire with 5-6 volutions and exactly 3 chambers per whorl, an arrangement which results in a very straight line of chambers when viewed from the side; chambers inflated, especially at apertural end; wall translucent, shiny, smooth, with fine perforations; sutures depressed; aperture of average size for genus.

LENGTH: range: 260-810; mean: 430; coefficient of variation; 0.28.

OBSERVATIONS: Our specimens lack the apical spine described and drawn by Fornasini.

DISTRIBUTION: see Figure 5.

Bulimina marginata d'Orbigny, Pl. 5, figs. 10-12.

ORIGINAL CITATION: *Bulimina marginata* d'Orbigny, 1826. Ann. Sci. Nat., sér. 1, v. 7, p. 269, no. 4, pl. 12, figs. 10, 12.

20

TYPE LOCALITY: Adriatic Sea near Rimini, Italy.

AGE: Probably Pliocene.

DESCRIPTION: Test variable in form, expanding rapidly from the apical to apertural end, the last few chambers occupying major part of test, step like in profile; spire with 4-5 volutions; chambers inflated, sub-triangular with angulate lower border which may be spinose or serrate; wall translucent and finely perforate; sutures depressed; aperture ovate.

LENGTH: range: 240-500; mean: 320; coefficient of variation; 0.19.

OBSERVATIONS: In Brazilian waters this species is covered with small spines whereas in the Argentine littoral zone the spinosity is greatly reduced (Pl. 5, figs. 11, 12).

DISTRIBUTION: see Figure 5. This species is found also in the backish waters of Lagoa dos Patos.

Bulimina patagonica d'Orbigny, sensu lato, Pl. 5, figs. 13-17.

ORIGINAL CITATION: *Bulimina patagonica* d'Orbigny, 1839. Voy. Amér. Mérid., v. 5, pt. 5, p. 50, pl. 1, figs. 8, 9.

TYPE LOCALITY: Bahía San Blas, Argentina.

AGE: Recent.

DESCRIPTION: Test elongate, tapering toward the spinose apical end; spire with 5 volutions; chambers inflated, sometimes with short spines along the lower border; wall shiny, white, translucent or opaque, finely perforate; sutures strongly depressed; aperture ovate, of medium size.

OBSERVATIONS: This species is distinguished from *B. marginata* by its greater elongation. The two formae which occur in the area are discussed below.

Bulimina patagonica d'Orbigny, forma typica, Pl. 5, figs. 13-15.

CHARACTERISTIC TRAITS: This forma has spines on the lower margins of the chambers.

LENGTH: range: 300-800; mean: 470; coefficient of variation: 0.30.

DISTRIBUTION: see Figure 5.

Bulimina patagonica d'Orbigny, forma glabra, Pl. 5, figs. 16, 17.

ORIGINAL CITATION: *Bulimina patagonica* var. glabra Cushman and Wickenden, 1929. U.S. Nat. Mus., Proc., v. 75, no. 9, p. 9, pl. 14, fig. 1.

TYPE LOCALITY: Cumberland Bay, Juan Fernandez Island.

AGE: Recent.

CHARACTERISTIC TRAITS: This forma lacks the spines of forma typica but is the same in all other respects.

LENGTH: range: 300-550; mean: 410; coefficient of variation: 0.15.

DISTRIBUTION: see Figure 5. The reduced ornamentation in forma glabra seems to be associated with its occurrence in brackish water (Río Quequén).

Bulimina cf. *B. pseudoaffinis* Kleinpell, Pl. 5, figs. 18, 19.

ORIGINAL CITATION: *Bulimina pseudoaffinis* Kleinpell, 1938. Miocene Stratigraphy of California. Amer. Assoc. Petrol. Geol., p. 257, pl. 9, fig. 9.

TYPE LOCALITY: Monterey Co., California.

AGE: Middle Miocene.

DESCRIPTION: Test conical, apertural end rounded due to width of last few chambers; chambers inflated with poorly developed spines on lower borders of early chambers; wall smooth, finely perforate; sutures depressed; aperture large, elongate.

LENGTH: range: 150-430.

OBSERVATIONS: Specimens from the southwest Atlantic are broader and smaller than those described by Kleinpell. These differences along with the presence of some poorly developed spines accounts for the designation as *B.* cf. *B. pseudoaffinis*. Some specimens are narrower at the apertural end and can be confused with *B. patagonica*, f. typica or with weakly ornamented specimens of *B. marginata*.

DISTRIBUTION: see Figure 5.

Bulimina pupoides d'Orbigny, Pl. 5, figs. 20, 21.

ORIGINAL CITATION: *Bulimina pupoides* d'Orbigny, 1846. Foram. Foss. Bass. Tert. Vienne, p. 185, pl. 11, figs. 11, 12.

TYPE LOCALITY: Nussdorf, Vienna.

AGE: Tertiary.

DESCRIPTION: Test cocoon shaped, periphery lobate; spire with 4 volutions; chambers inflated, rounded, final chamber with flat apertural face lying in plane of major axis; wall smooth, white, shiny, finely perforate; sutures depressed; aperture relatively wide.

LENGTH: range: 300-400.

DISTRIBUTION: see Figure 5.

Bulimina subulata Cushman and Parker, Pl. 6, figs. 1-3.

ORIGINAL CITATION: *Bulimina elongata* d'Orbigny, var. subulata Cushman and Parker, 1937. Cushman Lab. Foram. Res., Contr., v. 13, pt. 2, p. 51, pl. 7, figs. 6, 7.

TYPE LOCALITY: Baden, Vienna.

AGE: Miocene.

DESCRIPTION: Test elongate, sides parallel; spire with 5 volutions; chambers slightly inflated, last few considerably higher than earlier; wall smooth, finely perforate, covered with spines particularly on lower border of early chambers; sutures depressed; aperture elongate.

LENGTH: range: 310-480.

DISTRIBUTION: see Figure 5.

BULIMINELLA Cushman, 1911

Test free, elongate, early chambers triserially arranged, later spiral with few whorls and many chambers per whorl, high spired; chambers very narrow, elongate parallel to axis of coiling; wall calcareous, translucent, perforate, smooth or striate, rarely spinose, sutures obscure; aperture ovate with internal tooth plate in umbilicus of last whorl and connected with prior chamber.

Buliminella auricula (Heron-Allen and Earland), Pl. 6, figs. 4-6.

ORIGINAL CITATION: *Bulimina auricula* Heron-Allen and Earland, 1932. Discovery Repts., v. 4, p. 351, pl. 9, figs. 1, 2.

TYPE LOCALITY: Islas Malvinas.

AGE: Recent.

DESCRIPTION: Test with 2-2½ volutions, peripheral margin rounded; final whorl with 5 chambers which occupy all of one side; wall smooth, shiny, white, finely perforate; sutures wide, slightly depressed; ventral side with radial grooves running from periphery to central zone where they empty into the apertural cavity.

LENGTH: range: 146-220.

DISTRIBUTION: see Figure 5.

Buliminella elegantissima (d'Orbigny), Pl. 6, figs. 7-10.

ORIGINAL CITATION: *Bulimina elegantissima* d'Orbigny, 1839. Voy. Amér. Mérid., v. 5, pt. 5, p. 51, pl. 7, figs. 13, 14.

TYPE LOCALITY: West coast of South America.

AGE: Recent.

DESCRIPTION: Test fusiform, rounded at extremities; spire with 2-3 volutions; chambers elongate, curved, narrow and obliquely inclined to coiling axis, 7-9 in final whorl; wall smooth, shiny, finely perforate; sutures depressed; aperture small, round and located in slight depression in apertural face, surrounded by broad flange.

LENGTH: range: 210-400; mean: 300; coefficient of variation: 0.17.

OBSERVATIONS: This species cannot be found among the original d'Orbigny material.

DISTRIBUTION: see Figure 5. This species apparently can tolerate brackish water as it is found living in the eastern part of the Río de la Plata and at Arroio Chuí.

Buliminella seminuda (Terquem), Pl. 6, figs. 11-15.

ORIGINAL CITATION: *Bulimina semi-nuda* Terquem, 1882. Soc. Géol. France, Mem., sér. 3, v. 3, no. 3, p. 117, pl. 12, fig. 21.

TYPE LOCALITY: Vicinity of Paris.

AGE: Eocene.

DESCRIPTION: Test elongate, oval, apertural end in form of truncated cone, apical end rounded, central region inflated, periphery rounded; spire with 3-4 volutions; chambers elongate, obliquely inclined; wall smooth, shiny; sutures flush; aperture small with shallow grooves radiating from aperture to peripheral margin of apertural face.

LENGTH: range: 260-630; mean: 410; coefficient of variation: 0.22.

DISTRIBUTION: see Figure 13.

Cancris Montfort, 1808

Test trochospiral, elongate biconvex, spiral side evolute, umbilical side slightly open; chambers increase rapidly in size, wall calcareous, perforate except on part of umbilical side of ultimate chamber; aperture interiomarginal, narrow.

Cancris sagra (d'Orbigny), Pl. 6, figs. 16-18.

ORIGINAL CITATION: *Rotalina sagra* d'Orbigny, 1839. In: de la Sagra, Hist. Phys. Polit. Natur. Cuba, p. 91, pl. 5, figs. 13-15.

TYPE LOCALITY: Cuba and Jamaica.

AGE: Recent.

DESCRIPTION: Test oblong, umbilical side convex, spiral side nearly flat, peripheral margin carinate; spire with two volutions; chambers triangular on umbilical side, lunate on spiral side, 6 in final whorl, final chamber on umbilical side occupying almost half of test; wall transparent, densely perforate with fine pores; sutures arcuate, limbate, raised on spiral side, depressed on umbilical side; aperture a broad slit from periphery to umbilicus, covered by extension of last chamber.

LENGTH: range: 440-850; mean: 720; coefficient of variation: 0.18.

DISTRIBUTION: see Figure 5.

CASSIDULINA d'Orbigny, 1826

Test lenticular or subglobular, involute, commonly biumbonate; chambers arranged in a biserial manner on both sides of periphery; wall calcareous, hyaline, granular, perforate, smooth; aperture an elongate oval or slit extending from the basal suture and parallel to the margin of the penultimate chamber, with a narrow lip along the lower border and with a tooth plate.

Cassidulina crassa d'Orbigny, sensu lato, Pl. 6, figs. 19-23; Pl. 7, figs. 1-3.

ORIGINAL CITATION: *Cassidulina crassa* d'Orbigny, 1839. Voy. Amér. Mérid., v. 5, pt. 5, p. 56, pl. 7, figs. 18-20.

TYPE LOCALITY: Isles Malvinas and Cabo de Hornos.

AGE: Recent.

DESCRIPTION: Test oval to circular, peripheral margin broadly rounded; chambers slightly inflated, the last rectangular to trapezoidal; wall thick, smooth, shiny, white, finely perforate; sutures slightly depressed; aperture ovate, almost perpendicular to basal suture, flanked by a broad rim.

OBSERVATIONS: Assemblages of this species are composed of two distinct end members which intergrade completely with each other.

Cassidulina crassa d'Orbigny, forma typica, Pl. 6, figs. 19-23.

CHARACTERISTIC TRAITS: The test outline is circular and the final chamber rectangular.

DIAMETER: range: 450-1180; mean: 790; coefficient of variation: 0.22.

OBSERVATIONS: There are five well preserved examples of this species in the original collection of d'Orbigny. The size of the pores in this species is greater than that of either *C. minuta* or *C. rossensis*.

DISTRIBUTION: see Figures 5 and 13.

Cassidulina crassa d'Orbigny, forma porrecta, Pl. 7, figs. 1-3.

ORIGINAL CITATION: *Cassidulina crassa* d'Orbigny var. porrecta Heron-Allen and Earland, 1932. Discovery Repts., v. 4, p. 358, pl. 9, figs. 34-37.

TYPE LOCALITY: Islas Malvinas and Tierra del Fuego.

AGE: Recent.

CHARACTERISTIC TRAITS: This forma is distinguished from forma typica by its pyriform outline and trapezoidal final chamber.

DIAMETER: range: 540-1090; mean: 790; coefficient of variation: 0.22.

DISTRIBUTION: see Figures 5 and 13.

Cassidulina laevigata d'Orbigny, Pl. 7, figs. 4-6.

ORIGINAL CITATION: *Cassidulina laevigata* d'Orbigny, 1826. Ann. Sci. Nat., sér. 1, v. 7, p. 282, pl. 15, figs. 4, 5.

TYPE LOCALITY: Not given.

AGE: Not given.

DESCRIPTION: Test compressed, subcircular, peripheral margin carinate, final whorl with five elongate chamber pairs; wall thin, smooth, shiny, with medium sized pores rather widely scattered over the surface except in the central zone; sutures depressed, arcuate; aperture narrow, parallel to the periphery, with a well developed lip extending into the aperture from the basal suture.

DIAMETER: range: 130-360; mean: 230; coefficient of variation: 0.26.

DISTRIBUTION: see Figure 5.

Cassidulina minuta Cushman, Pl. 7, figs. 7-11.

ORIGINAL CITATION: *Cassidulina minuta* Cushman, 1933. Cushman Lab. Foram. Res., Contr., v. 9, pt. 4, p. 92, pl. 10, fig. 3.

TYPE LOCALITY: Paumotu Islands.

AGE: Recent.

DESCRIPTION: Test subcircular, compressed, peripheral margin rounded; chambers inflated, 4 pairs in final whorl, rectangular, arranged perpendicular to preceeding chamber; wall smooth, delicate, shiny; sutures depressed; aperture a narrow elongate arc, containing a dental plate rising from the basal suture.

DIAMETER: range: 160-330; mean: 240; coefficient of variation: 0.21.

OBSERVATIONS: The pore size of *C. minuta* is intermediate between that of *C. crassa* and *C. rossensis*. *C. minuta* is dinstinguished from *C. crassa* by its greater compression.
DISTRIBUTION: see Figures 5 and 13.

Cassidulina pulchella d'Orbigny, Pl. 7, figs. 12-14.
ORIGINAL CITATION: *Cassidulina pulchella* d'Orbigny, 1839. Voy. Amér. Mérid., v. 5, pt. 5, p. 57, pl. 8, figs. 1-3.
TYPE LOCALITY: San Gallan, Peru.
AGE: Recent.
DESCRIPTION: Test outline subcircular, periphery rounded; chambers flattened, six pairs in final whorl, the final one of which is rectangular on the side without the aperture; wall shiny, finely perforate; sutures distinct, only slightly depressed; aperture elongate, parallel to peripheral margin.
DIAMETER: range: 260-500; mean: 370; coefficient of variation: 0.14.
OBSERVATIONS: D'Orbigny's original figure of this species which shows a serrate border is incorrect. The four well preserved specimens in the Paris collection are circular in outline.
DISTRIBUTION: see Figures 5 and 13.

Cassidulina rossensis (Kennett), Pl. 7, figs. 15-17.
ORIGINAL CITATION: *Globocassidulina crassa* (d'Orbigny) *rossensis* Kennett, 1967. Cushman Found. Foram. Res., Contr., v. 18, pt. 3, p. 134, pl. 11, figs. 4-6.
TYPE LOCALITY: McMurdo Sound, Ross Sea.
AGE: Recent.
DESCRPTION: Test outline subcircular, compressed; chambers somewhat inflated, polygonal, 4-5 pairs in last volution; wall smooth, delicate, shiny, very finely perforate, white; sutures radial, depressed; aperture elongate and narrow, bounded by a distinct flange on the basal suture, with perpendicular branches at both ends, the uppermost of which is covered by a small dental plate.
DIAMETER: range: 200-470; mean: 320; coefficient of variation: 0.19.
DISTRIBUTION: see Figures 5 and 13.

Cassidulina subglobosa Brady, Pl. 7, figs. 18-20.
ORIGINAL CITATION: *Cassidulina subglobosa* Brady, 1881. Quart. Jour. Micr. Sci., v. 21, p. 60; 1884. Challenger Exp., Repts., Zool., v. 9, p. 430, pl. 54, fig. 17.
TYPE LOCALITY: Not designated (figured specimen from off Pernambuco, north Brazilian coast).
AGE: Recent.
DESCRIPTION: Test subglobular; chambers strongly inflated, disguising the biserial nature of coil, four pair in final whorl; wall white, shiny, very finely perforate; sutures depressed; aperture ovate and almost perpendicular to basal suture; apertural face with faint striae radiating outward from aperture.

DIAMETER: range: 110-600; mean: 250; coefficient of variation: 0.44.
DISTRIBUTION: see Figures 5 and 13.

CASSIDULINOIDES Cushman, 1927

Test free, elongate, biserial arrangement of chambers coiled in a planispiral fashion, involute and subglobular in early stages, evolute and rectilinear in later stages; wall calcareous, perforate; aperture ovate, terminal.

Cassidulinoides parkerianus (Brady), Pl. 8, figs. 1-4.
ORIGINAL CITATION: *Cassidulina parkeriana* Brady, 1884. Challenger Exp., Repts., Zool., v. 9, p. 432, pl. 54, figs. 11-16.
TYPE LOCALITY: Not designated (figured specimens from west coast of Patagonia).
AGE: Recent.
DESCRIPTION: Test J-shaped or subglobular, linear portion with 3-4 pairs of chambers lying perpendicular to the growth axis; chambers globose, quadrate; wall thin, white, opalescent, with numerous large pores; sutures depressed; aperture small, lying in a cavity atop or along side the apertural face.
LENGTH: range: 240-750; mean: 410; coefficient of variation: 0.39.
DISTRIBUTION: see Figure 13.

CIBICIDES Montfort, 1808

Test attached, trochospiral, spiral side planate to concave, evolute, umbilical side convex, involute, peripheral margin angulate with imperforate border; wall calcareous, radial, bilamellar, generally coarsely perforate on spiral side, finely perforate on umbilical side; aperture interiomarginal, extending across the peripheral margin and along the spiral suture as far as the fourth chamber, with lip.

Cibides ex gr. *C. aknerianus* (d'Orbigny), Pl. 8, figs. 5-11.
ORIGINAL CITATION: *Rotalina akneriana* d'Orbigny, 1846. Foram. Foss. Bass. Tert. Vienne, p. 156, pl. 8, figs. 13-15.
TYPE LOCALITY: Nussdorf, Vienna.
AGE: Tertiary.
DESCRIPTION: Test outline subcircular, lobate, peripheral margin subangular to carinate; umbilical side showing part of earlier whorls; spiral side with 2½-3 whorls, the last of which with 8-11 slightly inflated chambers; wall opaque, coarsely perforate on both sides; aperture a narrow slit with distinct lip.
DIAMETER: range: 260-850; mean: 500; coefficient of variation: 0.34.
OBSERVATIONS: This species varies considerably in size, number of chambers, nature of the peripheral margin and

23

degree of convexity and porosity. It is widely distributed over the shelf with the largest and best developed specimens occurring on the outer shelf in the subantarctic waters of the Malvin Current.

This species can be confused with specimens of *C. dispars* which lack the umbilical knob, but it is generally less rounded in outline and more compressed. It can also be confused with *C. variabilis* but is more massive than this species.

DISTRIBUTION: see Figures 6 and 13.

Cibicides dispars (d'Orbigny), Pl. 8, figs. 12-16.
ORIGINAL CITATION: *Truncatulina dispars* d'Orbigny, 1839. Voy. Amér. Mérid., v. 5, pt. 5, p. 38, pl. 5, figs. 25-27.
TYPE LOCALITY: Islas Malvinas.
AGE: Recent.
DESCRIPTION: Test outline ovate to subcircular, slightly lobate, peripheral margin subacute, chambers inflated, triangular, curved on umbilical side, quadrate and flat on spiral side, 7-10 in final whorl; umbilical knob may be present; wall smooth, white, opaque or translucent, with fine perforation on the umbilical side and large pores on spiral side; sutures obscure on early chambers of both sides; aperture visible along spiral suture of last 3-4 chambers with narrow lip.
DIAMETER: range: 270-600; mean: 450; coefficient of variation: 0.16.
OBSERVATIONS: This species is quite variable. Some specimens have a distinct umbilical knob while others do not. Some specimens possess larger pores on the spiral side than on the umbilical side whereas others are equally porous on both sides. The number in the final whorl varies between 7-10 although d'Orbigny describes only 8. The non-umbonate examples of this species are sometimes confused with *C. aknerianus*.
DISTRIBUTION: see Figures 6 and 13.

Cibicides cf. *C. fletcheri* Galloway and Wissler, Pl. 8, figs. 17-21.
ORIGINAL CITATION: *Cibicides fletcheri* Galloway and Wissler, 1927. Jour. Paleontol., v. 1, p. 64, pl. 10, figs. 8, 9.
TYPE LOCALITY: Lomita Quarry, California.
AGE: Pleistocene.
DESCRIPTION: Test outline ovate or subcircular, plano-convex, somewhat compressed, peripheral outline lobate and subacute, umbilical side with distinct circular umbilical knob and 9-11 chambers, spiral side showing 2½ whorls; chambers low and short in early part of whorl, expanded in later part; wall white to pale yellow, coarsely porous; aperture arcuate, with lip.
DIAMETER: range: 300-700; mean: 460; coefficient of variation: 0.22.
OBSERVATIONS: The majority of the specimens assigned to this taxon are not typical *C. fletcheri* since they do not

possess a greatly enlarged last chamber, since their outline is more circular and since their umbilical side is too convex. Moreover, the number of chambers in the final whorl is seldom 11 as in typical *C. fletcheri*, but is more commonly 9. Specimens corresponding to *C. fletcheri* are found only in some areas of the Malvin Current.
DISTRIBUTION: see Figures 6 and 13.

Cibicides lobatulus (Walker and Jacob) pl. 9, figs. 1-4.
ORIGINAL CITATION: *Nautilus lobatulus* Walker and Jacob, 1798, Ess. Micr., p. 642, pl. 14, fig. 36.
TYPE LOCALITY: Whitstable, Kent, England.
AGE: Recent.
DESCRIPTION: Test ovate in outline, markedly lobate, very compressed, umbilical side convex, spiral side planar or slightly concave, early whorls visible; peripheral margin acute or carinate; chambers inflated, 6-7 in final whorl; umbilicus depressed; wall on umbilical side coarsely perforate or lacking pores, on spiral side pores intermediate in size and uniformly distributed; sutures limbate on spiral side, depressed on umbilical side; aperture with distinct lip and extending rearward to the penultimate chamber along the spiral suture, also extending along the base of the apertural face on the umbilical side.
DIAMETER: range: 300-640.
DISTRIBUTION: see Figures 6 and 13.

Cibicides mckannai Galloway and Wissler, Pl. 9, figs. 5-8.
ORIGINAL CITATION: *Cibicides mckannai* Galloway and Wissler, 1927, Jour. Paleontol., v. 1, p. 65, pl. 10, figs. 5, 6.
TYPE LOCALITY: Lomita Quarry, California.
AGE: Pleistocene.
DESCRIPTION: Test outline subcircular, planoconvex to unequally biconvex, umbilical convexity due to well developed umbo, peripheral margin subacute; chambers low, curved, 9-10 in final whorl, on spiral side covered by thick calcareous material often forming a central knob; wall opaque, yellow to yellow-brown; sutures arcuate, thick, somewhat depressed on umbilical side, covered with clear material; aperture extending along spiral suture for four chambers.
DIAMETER: range: 210-440; mean: 310; coefficient of variation: 0.16.
OBSERVATIONS: This is a highly variable species, especially in its convexity and degree of porosity. Typical examples are found only in the Malvin Current. In parts of the inner shelf there are forms with fewer chambers, less convexity and more variable sutures that we would designate as *C. cf. C. mckannai* but they are rare and are therefore not included here.
DISTRIBUTION: see Figure 13.

Cibicides refulgens Montfort, Pl. 9, figs. 9-11.
ORIGINAL CITATION: *Cibicides refulgens* Montfort, 1808.

Conchyl. Syst., v. 1, p. 123, p. 122, textfig.

TYPE LOCALITY: Livorno and Sienna, Italy.

AGE: Not designated.

DESCRIPTION: Test outline subcircular, strongly plano-convex, peripheral margin angulate with keel; chambers low, 7-8 in final whorl; wall thin, finely and irregularly perforate on umbilical side, coarsely perforate on spiral side; all whorls visible on spiral side; sutures limbate on spiral side, depressed on umbilical side; aperture extending along entire spiral suture of last chamber and occupying one-third of basal suture on umbilical side.

DIAMETER: range: 260-380.

DISTRIBUTION: see Figure 6.

Cibicides variabilis (d'Orbigny), Pl. 9, figs. 12-17.

ORIGINAL CITATION: *Truncatulina variabilis* d'Orbigny, 1826, Ann. Sci., Nat., sér. 1, v. 7, p. 279. 1839. In: Barker-Webb and Berthelot, Hist. Nat. Iles Canaries, v. 5, pt. 2, p. 135, pl. 2, fig. 29.

TYPE LOCALITY: Mediterranean Sea.

AGE: Recent.

DESCRIPTION: Test variable in outline, compressed, peripheral margin acute, lobate with keel, umbilical side slightly convex, partially evolute, spiral side planar; chambers on umbilical side triangular in early whorls, later quadrangular, somewhat globose and sometimes irregularly distributed, even becoming biserial at times, 6-8 chambers in final whorl; wall coarsely perforate; sutures limbate on spiral side, depressed on umbilcal side; aperture extending along spiral suture for one or two chambers, with thick lip.

DIAMETER: range: 510-1290; mean: 760; coefficient of variation: 0.28.

OBSERVATIONS: The name of this species is justified as there is a great deal of variation in the disposition of the chambers, depending mainly on the character of the substrate where they live. Large populations, as are found south of Tierra del Fuego, exhibit all extremes of variation with complete transition between them. These forms could be interpreted as belonging to such genera as *Dyocibicides*, *Cibicidella*, and *Cyclocibicides*. We consider these genera to lack zoological validity.

Some examples of this species are similar to *C*. cf. *C. aknerianus*.

DISTRIBUTION: see Figures 6 and 14.

CRIBROROTALIA Hornibrook, 1961

Test free, trochospiral, biconvex; umbilical area filled with large knob containing anastomosing labyrinthic canals; wall calcareous, radial, finely perforate; aperture formed by one or more lines of pores at base of apertural face.

Cribrorotalia meridionalis (Cushman and Kellett), Pl. 9, figs. 18-20.

ORIGINAL CITATION: *Eponides meridionalis* Cushman and Kellett, 1929. U.S. Nat. Mus., Proc., v. 75, art. 25, p. 11, pl. 4, figs. 4-6.

TYPE LOCALITY: Corral, Chile.

AGE: Recent.

DESCRIPTION: Test circular in outline, peripheral margin acute or carinate; chambers triangular on umbilical side, 9-10 in final whorl, curved and quadrangular on spiral side; wall opaque, densely and finely perforate, covered with both large and small irregular distributed granules on umbilical side; sutures limbate and raised on both sides, slightly depressed near periphery on umbilical side; aperture a line of rimmed pores along base of apertural face.

DIAMETER: range: 380-920; mean: 670; coefficient of variation: 0.21.

DISTRIBUTION: see Figure 6.

CRIBROSTOMOIDES Cushman, 1910

Test free, planispiral, involute, wall agglutinate, simple, with abundant cementing agent; aperture interio-areal.

Cribrostomoides crassimargo (Norman), Pl. 10, figs. 1-3.

ORIGINAL CITATION: *Haplophragmium canariense* (d'Orbigny). Brady, 1884, Challenger Repts., Zool., v. 9, p. 310, pl. 35, fig. 4 [non 1-3,5]: *Haplophragmium crassimargo* Norman, 1892. Museum Normanianum, pts. 7, 8, p. 17.

TYPE LOCALITY: Dogger Bank, North Sea.

AGE: Recent.

DESCRIPTION: Test subcircular in outline, partially evolute, peripheral margin rounded and lobate, umbilicus depressed; chambers inflated, triangular, 7-9 in last whorl; wall thick, glassy, composed of fine material and incrusted with coarse grains, sutures depressed, radial, straight; aperture a radial equatorial slit of uniform height throughout its length, located in the lower half of apertural face, with lip.

DIAMETER: range: 440-1420; mean: 970; coefficient of variation: 0.27.

DISTRIBUTION: see Figure 14.

Cribrostomoides jeffreysii (Williamson), Pl. 10, figs. 4-7.

ORIGINAL CITATION: *Nonionina jeffreysii* Williamson, 1858. Foram. Gr. Brit., Ray Soc., p. 34, pl. 3, figs. 72, 73.

TYPE LOCALITY: British Isles.

AGE: Recent.

DESCRIPTION: Test subcircular in outline, lobate, peripheral margin rounded, compressed; chambers inflated, arranged in partially evolute planispiral, 7-9 in final whorl; wall smooth, delicate, composed of small well sorted quartz grains held together by a gray or yellow-brown cement; sutures depressed; apertural face convex with ovate aperture lying perpendicular to plane of coiling, located in the lower one-third of the apertural face, with thin well developed lip.

DIAMETER: range: 210-970; mean: 550; coefficient of variation: 0.33.

OBSERVATIONS: This species is quite variable in size. The larger specimens best show the partially evolute mode of coiling.

DISTRIBUTION: see Figures 6 and 14.

Cribrostomoides weddellensis (Earland), Pl. 10, figs. 8-10.
ORIGINAL CITATION: *Haplophragmoides weddellensis* Earland, 1936. Discovery Repts., v. 13, p. 33, pl. 1, figs. 15, 16.
TYPE LOCALITY: Weddell Sea.
AGE: Recent.
DESCRIPTION: Test rounded in outline, irregularly involute, massive, peripheral margin rounded; chambers inflated, 3-5 in final whorl; wall rugose, composed of poorly sorted large grains; sutures obscure; aperture small and barely visible slit.
DIAMETER: range: 210-450.
DISTRIBUTION: see Figures 6 and 14.

CYCLOGYRA Wood, 1842

Test free, circular, evolute; proloculus globular, second chamber tubular, undivided, planispirally enrolled; wall calcareous, imperforate, porcellaneous; aperture terminal, occupying entire end of second chamber.

Cyclogyra involvens (Reuss), Pl. 10, figs. 11, 12.
ORIGINAL CITATION: *Operculina involvens* Reuss, 1850. Denkschr. Akad. Wiss. Wien, v. 1, p. 370, pl. 46, fig. 20.
TYPE LOCATION: Baden, Vienna.
AGE: Tertiary.
DESCRIPTION: Test circular in outline, disc shaped, compressed, peripheral margin rounded; second chamber a gradually expanding tube wrapped 5-6 times about a planispiral axis of coiling; wall smooth, shiny, translucent, delicate, exhibiting growth lines; sutures distinct, slightly depressed; aperture comprising entire terminal end of second (last) chamber.
DIAMETER: range: 110-500; mean: 270; coefficient of variation: 0.37.
DISTRIBUTION: see Figures 6 and 14. This species is also found in the mixohaline waters of Río Quequén.

Cyclogyra planorbis (Schultze), Pl. 10, figs. 13-15.
ORIGINAL CITATION: *Cornuspira planorbis* Schultze, 1854. Organ. Polythal., Engelmann, p. 40, pl. 2, fig. 21.
TYPE LOCALITY: Mozambique coast.
AGE: Recent.

DESCRIPTION: Test compressed, slightly asymmetrical in that final volution exhibits trochospiral coiling; second chamber with 3-4 whorls; peripheral margin rounded; sides concave due to marked expansion of diameter of second chamber; wall fragile, translucent, exhibiting growth lines; sutures distinctly depressed; aperture large.

DIAMETER: range: 100-380; mean: 240; coefficient of variation: 0.33.

OBSERVATIONS: This species differs from *C. involvens* by having fewer volutions and a second chamber whose diameter increases rapidly.

DISTRIBUTION: See Figure 6.

DAHLGRENIA Lena, 1974

Test free, unilocular, ovate or fusiform; transverse section circular or ovate; wall bilamellar, the external layer hyaline, gray, granular, composed of diatom frustules, the internal layer with entosolenian tube joined to internal layer by two lateral bands, giving aperture a biradial symmetry.

Dahlgrenia patagoniensis Lena, Pl. 10, Figs. 16-18.
ORIGINAL CITATION: *Dahlgrenia patagoniensis* Lena, 1974. Physis. Sec. A, v. 33, no. 86, p. 9, pls. 1-4.
TYPE LOCALITY: Puerto Deseado, Argentina.
AGE: Recent.
DESCRIPTION: As this species is monotypic, its description is that of the genus.
LENGTH: range: 240-540; mean: 350; coefficient of variation: 0.20.
DISTRIBUTION: see Figure 6.

DENTALINA Risso, 1826

Test elongate, uniserial, arcuate; sutures commonly oblique; aperture radiate, terminal, central to marginal.

Dentalina communis (d'Orbigny), Pl. 10, figs. 19, 20.
ORIGINAL CITATION: *Nodosaria (Dentalina) communis* d'Orbigny, 1826. Ann. Sci. Nat., sér. 1, v. 7, p. 254; 1865. Parker and Jones, Phil. Trans. Roy. Soc., v. 155, no. 1, p. 342, pl. 13, fig. 10.
TYPE LOCALITY: Adriatic Sea.
AGE: Recent.
DESCRIPTION: Test long, transverse section circular; chambers slightly inflated, increasing in diameter toward apertural end, 4-9 in number; wall thick, finely perforate, often translucent; sutures slightly depressed, oblique; aperture offset toward the concave margin.
LENGTH: range: 1460-2840.
DISTRIBUTION: see Figures 6 and 14.

DISCORBINELLA Cushman and Martin, 1935

Test free, spiral side convex, partially evolute, displaying part of previous whorl, umbilical side planate to slightly concave, involute, peripheral margin carinate or keeled; umbilical side chambers with strongly developed internal flap; wall calcareous, perforate, radial, monolamellar; sutures distinct; aperture interiomarginal, peripheral, with supplementary apertures along margins of chamber flaps.

Discorbinella altocamerata (Heron-Allen and Earland), Pl. 11, figs. 1-4.

ORIGINAL CITATION: *Truncatulina tenuimargo* Brady var. altocamerata Heron-Allen and Earland, 1922. Brit. Antarct. Exp., Brit. Mus. (Nat. Hist.), Zool., v. 6, no. 2, pt. 2, p. 209, pl. 7, figs. 24-27.

TYPE LOCALITY: Not designated.

AGE: Not designated.

DESCRIPTION: Test subcircular in outline, umbilical side convex, spiral side planar or slightly convex, peripheral margin sharply carinate or keeled; chambers conical, on umbilical side lunate, 7-8 in final whorl; wall opaque to translucent, finely perforate; sutures limbate, curved, on umbilical side flush, on spiral side strongly depressed; aperture small, ovate, with rim, peripheral, at base of last chamber; 2-4 supplementary apertures at margin of flaps which extend into umbilicus from each chamber.

DIAMETER: range: 260-530.

DISTRIBUTION: see Figures 6 and 14.

DISCORBIS Lamarck, 1804

Test free or adhered, trochospiral; chambers numerous with flap extending into umbilicus beneath which is a cavity connecting with umbilicus; wall calcareous, perforate; sutures depressed or limbate; principal aperture an interiomarginal slit; secondary apertures below umbilical flaps.

Discorbis bertheloti (d'Orbigny), Pl. 11, figs. 5-7.

ORIGINAL CITATION: *Rosalina bertheloti* d'Orbigny, 1839. In: Barker-Webb and Berthelot, Hist. Nat. Iles Canaries, p. 135, pl. 1, figs. 28-30.

TYPE LOCALITY: Teneriffe, Canary Islands.

AGE: Recent.

DESCRIPTION: Test compressed, oval in outline, spiral side convex, umbilical side concave and partially evolute, sometimes revealing the proluculus, peripheral margin carinate; chambers arcuate, 6-7 in final whorl; wall translucent, densely perforate, shiny; sutures curved, depressed; aperture narrow, extending to periphery; chamber flaps extending into umbilicus are weakly developed and often absent.

DIAMETER: range: 170-280.

DISTRIBUTION: see Figure 6.

Discorbis isabelleanus (d'Orbigny), Pl. 11, figs. 8-12.

ORIGINAL CITATION: *Rosalina isabelleana* d'Orbigny, 1839. Voy. Amér. Mérid., v. 5, pt. 5, p. 43, pl. 6, figs. 10-12.

TYPE LOCALITY: Islas Malvinas.

AGE: Recent.

DESCRIPTION: Test ovate, equally biconvex, peripheral margin acute and carinate; chambers elongate and obliquely arranged on spiral side, triangular on umbilical side except final chamber which is hemispherical, 5-7 in final whorl; wall thick, finely perforate and granular on spiral side, smooth and shiny on umbilical side; sutures limbate and often raised above surface on spiral side, depressed on umbilical side; aperture narrow, long, bordered by lip.

DIAMETER: range: 200-1630; mean: 890; coefficient of variation: 0.57.

DISTRIBUTION: see Figure 14.

Discorbis malovensis Heron-Allen and Earland, Pl. 11, figs. 13-15.

ORIGINAL CITATION: *Discorbis malovensis* Heron-Allen and Earland, 1932. Discovery Repts., v. 4, p. 415, pl. 14, figs. 22-24.

TYPE LOCALITY: Islas Malvinas.

AGE: Recent.

DESCRIPTION: Test conical, circular in outline, peripheral margin subacute, spiral side convex exhibiting 3-4 volutions, each with 4 (rarely 5) chambers which are narrow and curved; wall pale yellow to glassy white, finely perforate, glassy on spiral side, granulate in central zone on umbilical side; sutures on spiral side flush, on umbilical side obscure; aperture a curved slit.

DIAMETER: range: 380-440.

OBSERVATIONS: The sutures are indistinguishable under the SEM.

DISTRIBUTION: see Figure 14.

Discorbis peruvianus (d'Orbigny), Pl. 11, figs. 16-20.

ORIGINAL CITATION: *Rosalina peruviana* d'Orbigny, 1839. Voy. Amér. Mérid., v. 5, pt. 5, p. 41, pl. 1, figs. 12-14.

TYPE LOCALITY: Coast of Peru and Chile.

AGE: Recent.

DESCRIPTION: Test outline irregularly ovate and lobate, spiral side convex with 2½-3 whorls, umbilical side concave with umbilicus only partially covered by cameral flaps; chambers triangular to quadrangular, arcuate, 5-6 in final whorl; wall irregularly perforate, yellow to orange; sutures on spiral side limbate in early chambers, depressed on later chambers, on umbilical side depressed; aperture extends from umbilicus to periphery.

DIAMETER: range: 270-580; mean: 380; coefficient of variation: 0.21.

27

OBSERVATIONS: This species exhibits considerable variation in the nature of the sutures, test shape and chamber shape. Comparison of the forms from the western coast of South America (topotype material?) with the primary types of *D. floridana* (a widely known name) of the U.S. National Museum shows that the two are the same.

DISTRIBUTION: see Figures 6 and 14. This species is also found in the brackish waters of the Río de la Plata, Mar Chiquita and Río Quequén.

Discorbis cf. *D. valvulatus* (d'Orbigny), Pl. 12, figs. 1-4.
ORIGINAL CITATION: *Rosalina valvulata* d'Orbigny, 1826. Ann. Sci. Nat., sér. 1, v. 7, p. 271. 1839, d'Orbigny, In: de la Sagra, Hist. Phys. Polit, Nat. Cuba, p. 96, pl. 3, figs. 21-23.
TYPE LOCALITY: Cuba, Jamaica and Martinique.
AGE: Recent.
DESCRIPTION: Test outline ovate to circular, peripheral margin acute to carinate, test compressed, $2\frac{1}{2}$-3 volutions, spiral side convex, umbilical side almost planar and only partially involute; chambers elongate, arcuate and arranged obliquely, 5-6 in final whorl; wall smooth and shiny on spiral side, covered with pustules in central zone of umbilical side; sutures flush on spiral side, strongly lobate on umbilical side; aperture elongate, sinuous, bounded by irregular cameral flap.
DIAMETER: range: 160-330; mean: 230; coefficient of variation: 0.21.
OBSERVATIONS: These specimens differ from *D. valvulatus* due to the presence of the umbilical pustules which obscure the characteristic configuration of the sutures.
DISTRIBUTION: see Figure 6.

Discorbis williamsoni (Chapman and Parr), sensu lato, Pl. 12, figs. 5-12.
ORIGINAL CITATION: *Rotalina nitida* Williamson, 1858. Foram. Gr. Brit., Ray Soc., p. 54, pl. 4, figs. 106-108 [emend. *Discorbis williamsoni* Chapman and Parr, 1932, Roy. Soc. Victoria, Proc., v. 44 (n.s.), p. 226]: *Discorbina praegeri* Heron-Allen and Earland, 1913, Roy. Irish Acad., Proc., v. 31, sec. 3, p. 122, pl. 10, figs. 8-10.
TYPE LOCALITY: British Isles.
AGE: Recent.
DESCRIPTION: Test circular in outline, plano-convex, peripheral margin carinate, subacute, spiral side with 3 volutions; chambers curved and quadrate on spiral side, triangular with rounded umbilical margin on umbilical side, $6-8\frac{1}{2}$ in final whorl; umbilicus often filled with a knob whose size varies; wall translucent with pores which are irregularly distributed or absent; sutures limbate and very pronounced on final whorl on spiral side, depressed on umbilical side; aperture interiomarginal slit with lip.
DIAMETER: range: 210-470; mean: 330; coefficient of variation: 0.21.

OBSERVATIONS: A complete intergradation of forms between *D. williamsoni* and *D. praegeri* in many localities indicates that they should be placed in the same species.

The great variation in the nature of the umbilical knob (see Pl. 12, figs. 7, 9-11) casts doubt on the validity of the genus *Gavelinopsis* whose differences form *Discorbis* lie solely in the presence of this knob.

DISTRIBUTION: see Figures 6 and 14. This is a widespread and rather abundant species. Specimens from subantarctic waters are large and well developed (as is the case with *Cibicides* cf. *C. aknerianus*) whereas those of the inner shelf are considerably smaller. This species is also found in the brackish waters of Lagoa dos Patos and Arroio Chuí.

EHRENBERGINA Reuss, 1850

Test free, elongate, triangular in transverse section, planispiral in early portion, later unrolled, peripheral margin carinate, often with spines in various stages of development; chambers wide, low, biserially arranged; wall calcareous, granular, finely perforate, smooth or pustulose; aperture elongate, curved, parallel to periphery, with bordering lip.

Ehrenbergina pupa (d'Orbigny), Pl. 12, figs. 13-15.
ORIGINAL CITATION: *Cassidulina pupa* d'Orbigny, 1839. Voy. Amér. Mérid., v. 5, pt. 5, p. 57, pl. 7, figs. 21-23.
TYPE LOCALITY: Islas Malvinas.
AGE: Recent.
DESCRIPTION: Test elongate, triangular in cross-section, early chambers planispiral, later uncoiling and broadly expanding, peripheral margin subangulate in uncoiled section; chambers broad, low; wall smooth, finely perforate; sutures limbate, curved and distinct at apertural end; aperture elongate, curved, parallel to basal suture, with lip.
LENGTH: range: 330-700; mean: 450; coefficient of variation: 0.20.
DISTRIBUTION: see Figure 14.

ELPHIDIUM Montfort, 1808

Test free, planispiral, equally biconvex, involute, lenticular; chambers numerous with retral processes and interseptal bridges; retral processes opening into a complex internal canal system which extends between septa and chambers; sutures lined with rows of fossettes that form the external orifices of transverse canals; wall calcareous, radial or granulate.
OBSERVATIONS: Recent detailed study of the internal structure of *Elphidium* (Hansen and Lykke-Andersen, 1976) clarified much of the earlier confusion about this genus.

Elphidium advenum depressulum Cushman, Pl. 12, figs. 16-18.

ORIGINAL CITATION: *Elphidium advenum* (Cushman) var. *depressulum* Cushman, 1933, U.S. Nat. Mus., Bull. 161, pt. 2, p. 51, pl. 12, figs. a, b.

TYPE LOCALITY: Tonga Islands.

AGE: Recent.

DESCRIPTION: Test subcircular in outline, compessed, peripheral margin carinate, sometimes lobate; chambers slightly inflated, 9-11 in final whorl; umbilical zone with knob surrounded by small pustules which also occupy sutural fossettes; wall smooth, translucent, finely perforate; sutures depressed, narrow; fossettes variable; aperture a series of pores at base of apertural face.

DIAMETER: range: 170-370.

OBSERVATIONS: This is a highly variable species. Cushman considered the absence of an umbilical knob to be a distinguishing characteristic of *E. advenum* var. *depressulum*. However, we do not consider this of great importance since our specimens may or may not have this knob. Of greater taxonomic value is the observation that our specimens have fewer chambers in the final whorl and a more lobate and carinate peripheral margin that *E. advenum* and consequently we designated these specimens as *E. advenum depressulum*.

DISTRIBUTION: see Figure 6. This species is also found in the mixohaline waters of the Río de la Plata.

Elphidium alvarezianum (d'Orbigny), Pl. 12, figs. 19-21.

ORIGINAL CITATION: *Polystomella alvareziana* d'Orbigny, 1839. Voy. Amér. Mérid., v. 5, pt. 5, p. 31, pl. 3, figs. 11, 12.

TYPE LOCALITY: Patagonia and Islas Malvinas.

AGE: Recent.

DESCRIPTION: Test subcircular in outline, compressed, peripheral margin carinate; chambers low, arcuate, 8-12 in final whorl; wall white, opaque, thick; fossettes ovate, relatively large, surrounded by pustules; sutures broad, depressed; apertural face triangular, flat; aperture interiomarginal.

DIAMETER: range: 190-280.

DISTRIBUTION: see Figure 6 and 14.

Elphidium articulatum (d'Orbigny), Pl. 13, figs. 1-4.

ORIGINAL CITATION: *Polystomella articulata* d'Orbigny, 1839, Voy. Amér. Mérid., v. 5, pt. 5, p. 30, pl. 3, figs. 9, 10.

TYPE LOCALITY: Patagonia and Islas Malvinas.

AGE: Recent.

DESCRIPTION: Test subcircular in outline, lobate, peripheral margin rounded; chambers slightly inflated, broad, somewhat arcuate, 10-12 in final whorl; wall translucent or opaque, shiny, finely perforate; sutures relatively narrow with quadrangular fossettes and small pustules; apertural face convex; aperture a narrow slit at base of apertural face surrounded by tuberculate ornamentation.

DIAMETER: range: 240-580; mean: 370; coefficient of variation: 0.24.

OBSERVATIONS: Our specimens have an interiomarginal aperture with pustules whereas d'Orbigny described orifices in various parts of the apertural face. Moreover, our specimens have a peripheral margin that is more rounded than those described by d'Orbigny.

DISTRIBUTION: see Figures 7 and 14. This species also occurs in the brackish waters of Mar Chiquita.

Elphidium discoidale (d'Orbigny), Pl. 13, figs. 5-7.

ORIGINAL CITATION: *Polystomella discoidalis* d'Orbigny, 1839. In: de la Sagra, Hist. Phys. Polit. Natur. Cuba, p 76, pl. 6, figs. 23, 24.

TYPE LOCALITY: Not designated.

AGE: Recent.

DESCRIPTION: Test discoidal, peripheral margin subacute and thickened; chambers slightly inflated, arcuate, 8-12 in final whorl; wall shiny, delicate, smooth, with relatively large pores; umbilical area relatively large, convex with holes of various sizes; sutures slightly depressed with small ovoid fossettes; aperture consists of several round holes in central and basal portion of apertural face.

DIAMETER: range: 270-780; mean: 480; coefficient of variation: 0.29.

DISTRIBUTION: see Figure 7. This species also occurs in the brackish waters of numerous lagoons and estuaries in the northern part of the study area (Lagoa Mirim, Lagoa dos Patos, Arroio Chuí, estuary of the Río de la Plata, and Mar Chiquita).

Elphidium excavatum (Terquem), Pl. 13, figs. 8-11.

ORIGINAL CITATION: *Polystomella excavata* Terquem, 1876. Soc. Dunkerquoise, Mém., v. 19, p. 429, pl. 2, fig. 2.

TYPE LOCALITY: Beach at Dunkirk, France.

AGE: Recent.

DESCRIPTION: Test circular in outline, somewhat compressed, peripheral margin rounded and slightly lobate; chambers inflated, 7-10 in final whorl; wall smooth, transparent or opaque, radial; sutures gently curved, depressed, often covered with pustules; fossettes of variable size and shape; umbilici concave, covered with pustules; aperture formed by series of holes at base of smooth apertural face.

DIAMETER: range: 190-360: mean: 240; coefficient of variation: 0.17.

OBSERVATIONS: There is a great deal of variation both in the size and shape of the fossettes as well as the degree of umbilical granulation. This species is very similar to *E. incertum* and can be easily confused with it.

DISTRIBUTION: see Figure 7. This species occurs in the brackish waters of Lagoa dos Patos and Río Quequén.

Elphidium galvestonense Kornfeld, Pl. 13, figs. 12-14.

ORIGINAL CITATION: *Elphidium gunteri* Cole, var. *galvestonensis* Kornfeld, 1931. Stanford Univ., Geol., Contr., v. 1, p. 87, pl. 15, figs. 1-3.

TYPE LOCALITY: Galveston Island, Texas, USA.

AGE: Recent.

DESCRIPTION: Test subcircular in outline, compressed, peripheral margin rounded; chambers narrow, curved, 11-15 in final whorl; umbilical region covered with plug, often subdivided; wall smooth, shiny , translucent, finely perforate; sutures curved, often pustulate and bearing small fossettes; aperture interiomarginal, covered with pustules.

DIAMETER: range: 210-440; mean: 320; coefficient of variation: 0.16.

DISTRIBUTION: see Figure 7. This species is sometimes found in brackish waters, e.g., Lagoa dos Patos.

Elphidium gunteri Cole, Pl. 13, figs. 15-18.

ORIGINAL CITATION: *Elphidium gunteri* Cole, 1931. Florida Geol. Surv., Bull. 6, p. 34, p. 4, figs. 9, 10.

TYPE LOCALITY: Orange City, Florida, USA.

AGE: Recent.

DESCRIPTION: Test subcircular to ovate in outline, expanded in umbilical area, peripheral margin broadly rounded; chambers slightly inflated, 10-13 in final whorl; wall radial, translucent to opaque, with large pores; sutures radial, deeply depressed with numerous fossettes lined with pustules; retral processes thick and distinct; umbilici with one or several knobs; aperture composed of various holes on and at base of apertural face, bordered by distinct rims.

DIAMETER: range: 200-500; mean: 290; coefficient of variation: 0.24.

DISTRIBUTION: see Figure 7. This species occurs also in the brackish waters of Lagoa dos Patos and Río Quequén.

Elphidium lessonii (d'Orbigny), Pl. 13, figs. 19, 20.

ORIGINAL CITATION: *Polystomella lessonii* d'Orbigny, 1826. Ann. Sci. Nat., sér. 1, v. 7, p. 284, no. 6. 1839. Voy. Amér. Mérid., v. 5, pt. 5, p. 29, pl. 3, figs. 1, 2.

TYPE LOCALITY: Patagonia and Islas Malvinas.

AGE: Recent.

DESCRIPTION: Test subcircular in outline, compressed, peripheral margin rounded; chambers low, broad, strongly arched, 14-18 in final whorl; wall translucent; sutures broad with 10-13 rectangular fossettes lined with tiny pustules; apertural face flat, lower portion covered with small holes each with fine lip.

DIAMETER: range: 280-1080; mean: 520; coefficient of variation: 0.37.

DISTRIBUTION: see Figures 7 and 14.

Elphidium macellum (Fichtel and Moll), sensu lato, Pl. 14, figs. 1-6.

ORIGINAL CITATION: *Nautilus macellus* Fichtel and Moll, 1798. *Testac. Microsc.*, p. 66, var. α, pl. 10, figs. e-g; var. β, pl. 10, figs. h-k.

TYPE LOCALITY: Mediterranean Sea.

AGE: Recent.

DESCRIPTION: Test lenticular in outline, peripheral margin carinate to keeled; chambers very low, raised, curved, 17-21 in final whorl; wall radial, translucent, shiny, sutures broad, with 10 or more rectangular and narrow fossettes lined with pustules (Pl. 14, fig. 4); umbilical zone somewhat convex with boss or several nodes; apertural face triangular, covered with small pustules; aperture multiple, interiomarginal, surrounded by small rims.

OBSERVATIONS: Populations of *E. macellum* exhibit two formae which intergrade with each other.

Elphidium macellum (Fichtel and Moll), forma typica, Pl. 14, figs. 1-4.

CHARACTERISTIC TRAITS: This forma has a carinate periphery, multiple nodes and many chambers in the final whorl.

DIAMETER: range: 330-910; mean: 580; coefficient of variation: 0.28.

DISTRIBUTION: see Figure 7 and 14.

Elphidium macellum (Fichtel and Moll), forma oweniana, Pl. 14, figs. 5, 6.

ORIGINAL CITATION: *Polystomella oweniana* d'Orbigny, 1839. Voy. Amér. Mérid., v. 5, pt. 5, p. 30, pl. 3, figs. 3, 4.

TYPE LOCALITY: Patagonian coast south of Río Negro.

AGE: Recent.

CHARACTERISTIC TRAITS: This forma is distinguished from forma typica by its well developed umbilical knob, keeled periphery and fewer chambers in the final whorl.

DIAMETER: range: 450-770.

DISTRIBUTION: see Figures 7 and 14.

Elphidium magellanicum Heron-Allen and Earland, Pl. 14, figs. 7-10.

ORIGINAL CITATION: *Elphidium (Polystomella) magellanicum* Heron-Allen and Earland, 1932. Discovery Repts., p. 440, pl. 16, figs. 26-28.

TYPE LOCALITY: Tierra del Fuego and Estrecho de Magallanes.

AGE: Recent.

DESCRIPTION: Test subcircular in outline, distinctly lobate, peripheral margin rounded, sometimes asymmetrical; chambers broad, inflated, slightly curved, 6-7 in final whorl; wall radial, transparent to translucent, finely perforate, delicate, generally covered with small rounded pustules; sutures radial, depressed, narrow, with few small fossettes; umbilici concave; aperture a narrow slit.

DIAMETER: range: 240-430; mean: 320; coefficient of variation: 0.16.

DISTRIBUTION: see Figure 7.

Elphidium margaritaceum Cushman, Pl. 14, figs. 11-13.

ORIGINAL CITATION: *Elphidium advenum* (Cushman), var. margaritaceum Cushman, 1930. U.S. Nat. Mus., Bull. 104, pt. 7, p. 25, pl. 10, fig. 3.

TYPE LOCALITY: Newport, Rhode Island, USA.

AGE: Recent.

DESCRIPTION: Test semicircular to ovate in outline, compressed, lobate, peripheral margin carinate; chambers arcuate, 8-10 in final whorl; wall translucent or opaque, radial, covered with pustules; sutures depressed; umbilici concave; aperture formed by interiomarginal holes, almost completely covered by pustules.

DIAMETER: range: 170-310.

DISTRIBUTION: see Figure 7.

EPISTOMINELLA Musezima and Maruhasi, 1944

Test free, trochospiral, biconvex or plano-convex; spiral side evolute, umbilical side involute with 6-7 chambers; wall calcareous, delicate, finely perforate; sutures clear, oblique on spiral side, radial on umbilical side; aperture a slit parallel to peripheral margin.

Epistominella exigua (Brady), Pl. 14, figs. 14-17.

ORIGINAL CITATION: *Pulvinulina exigua* Brady, 1884. Challenger Exped., Repts., Zool., v. 9, p. 696, pl. 103, figs. 13, 14.

TYPE LOCALITY: Not designated (probably Pacific Ocean).

AGE: Recent.

DESCRIPTION: Test subcircular in outline, slightly lobate, peripheral margin subacute, biconvex; chambers very slightly inflated, 5-6 in final whorl; wall transparent to translucent; sutures slightly arcuate, tangential and limbate on spiral side, depressed on umbilical side; aperture large, parallel to peripheral margin, wider at base where it butts against chamber margin.

DIAMETER: range: 130-240; mean: 190; coefficient of variation: 0.11.

OBSERVATIONS: This species, apart from being quite abundant, exhibits a great deal of variability in outline, size, convexity and nature of the sutures.

DISTRIBUTION: see Figures 7 and 14. This is a rather widespread species which also lives in the brackish waters of the Río de la Plata and Río Quequén.

FISSURINA Reuss, 1850

Test free, unilocular ovate, pyriform or rounded, bilaterally symmetrical, ovate in longitudinal section, elliptical, triangular or tetragonal in transverse section, with or without peripheral spines; wall hyaline, perforate, smooth or costate, reticulate, or foveolate; aperture an ovate or rounded fissure, terminal, may be located at end of prolongation of test, may have entoselenian tube extending from aperture into chamber.

Fissurina auriculata (Brady), Pl. 14, figs. 18-20.

ORIGINAL CITATION: *Lagena auriculata* Brady, 1881. Quart.

Jour. Micro. Sci., n.s., v. 21, p. 61; 1884. Challenger Expd., Repts., Zool., v. 9, pl. 60, fig. 31 (non 29, 33).

TYPE LOCALITY: Not designated.

AGE: Recent.

DESCRIPTION: Test subtriangular in outline, somewhat compressed at the apertural end and inflated at aboral end, peripheral margin subacute with an ovate depression bordered by alae located along both sides of aboral margin; wall smooth, translucent, finely perforate; aperture fusiform, elevated, with slight thickening of wall.

LENGTH: range: 210-330.

DISTRIBUTION: see Figure 14.

Fissurina bisulcata (Heron-Allen and Earland), Pl. 15, figs. 1-3.

ORIGINAL CITATION: *Lagena bisulcata* Heron-Allen and Earland, 1932. Discovery Repts., v. 4, p. 380, pl. 9, figs. 29-32.

TYPE LOCALITY: East of Cabo Virgenes or Banco Burdwood, 46°30'-47°30'S.

AGE: Recent.

DESCRIPTION: Test subcircular in outline, peripheral margin thickened to form rounded flange, lateral faces convex, peripheral margins of aboral region bordered by deep grooves partially covered by delicate rim; wall translucent or opaque, smooth, finely perforate; aperture, fusiform, short.

LENGTH: range: 230-360.

DISTRIBUTION: see Figure 14.

Fissurina compressa (d'Orbigny), Pl. 15, figs. 4-6.

ORIGINAL CITATION: *Oolina compressa* d'Orbigny, 1839. Voy. Amér. Mérid., v. 5, pt. 5, p. 18, pl. 5, figs. 1, 2.

TYPE LOCALITY: Patagonia and Islas Malvinas.

AGE: Recent.

DESCRIPTION: Test ovate in outline, compressed, peripheral margin rounded; wall smooth, translucent, shiny, finely perforate; aperture fusiform with short entosolenian tube.

LENGTH: range: 140-240.

DISTRIBUTION: see Figures 7 and 15.

Fissurina aff. *F. earlandi* Parr, Pl. 15, figs. 7-10.

ORIGINAL CITATION: *Fissurina earlandi* Parr, 1950. B.A.N.Z. Antarct. Res. Exp. 1929-1931, Repts. ser. B, v. 5, pt. 6, p. 306, pl. 8, fig. 8.

TYPE LOCALITY: Indian Ocean, 66°10'S, 49°41'E.

AGE: Recent.

DESCRIPTION: Test ovate in outline, compressed, pyriform, peripheral margin with rim; wall translucent, shiny, pitted, with fine pores and small irregularly shaped slits covering the surface (Pl. 15, fig. 10); aperture large, wide, acuminate with smooth narrow rim; entosolenian tube curved and attached at its distal end to wall.

LENGTH: range: 170-330; mean: 210; coefficient of variation: 0.19.

OBSERVATIONS: Our specimens are designated as *F.* aff. *F. earlandi* as they differ from the specimens of Parr by having a longer entosolenian tube, more rounded outline, and pores in the central area of the sides.

DISTRIBUTION: see Figures 7 and 15. This species is also found in the brackish waters of Río Quequén.

Fissurina cf. *F. ellipica* Sequenza, Pl. 15, figs. 11-13.

ORIGINAL CITATION: *Fissurina elliptica* Sequenza, 1862. Descrip. Foram. Monotal. Messina, p. 60, pl. 2, fig. 3.

TYPE LOCALITY: Messina, Sicily.

AGE: Teritiary.

DESCRIPTION: Test ellipsoidal, compressed, sides parallel, peripheral margin rounded; wall shiny, smooth; aperture narrow, short, with long entosolenian tube.

LENGTH: range: 200-240.

OBSERVATIONS: As our specimens lack the carinate margin of *F. elliptica* as described by Sequenze, we designate it *F.* cf. *F. elliptica*.

DISTRIBUTION: see Figure 7.

Fissurina laevigata Reuss, Pl. 15, fig. 14-16.

ORIGINAL CITATION: *Fissurina laevigata* Reuss, 1850. K. Akad. Wiss. Wien, Math. Nat. Cl., Denkschr., v. 1, p. 366, pl. 46, fig. 1.

TYPE LOCALITY: Grinzing, Vienna.

AGE: Tertiary.

DESCRIPTION: Test ovate to pyriform in outline, inflated in aboral half, more compressed in apertural half, peripheral margin rounded, sometimes carinate; wall transparent to translucent, smooth, shiny; apertural area thickened, glassy, lunate; aperture long, margins parallel, with short entosolenian tube.

LENGTH: range: 170-270; mean: 210; coefficient of variation: 0.14.

OBSERVATIONS: The specimens from the southwest Atlantic differ from the original description by being less acuminate in the apertural region and by the occasional presence of short processes on the aboral margin. For a more detailed taxonomy see Boltovskoy (1954a).

DISTRIBUTION: see Figure 7 and 15.

Fissurina laureata (Heron-Allen and Earland), Pl. 15, figs. 21-23.

ORIGINAL CITATION: *Lagena laureata* (Heron-Allen and Earland, 1932. Discovery Repts., v. 4, p. 382, pl. 13, fig. 4.

TYPE LOCALITY: Islas Malvinas and areas near the Estrecho de Magallanes.

AGE: Recent.

DESCRIPTION: Test subcircular to ovate in outline, slightly inflated in center, peripheral margin keeled; wall irregularly costate along margin and honeycombed in center; aperture a long slit with bordering rim.

LENGTH: range: 170-210.

DISTRIBUTION: see Figures 7 and 15.

Fissurina lucida (Williamson), Pl. 15, figs. 17-20.

ORIGINAL CITATION: *Entosolenia marginata*, var. lucida Williamson, 1884. Ann. Mag. Nat. Hist., ser. 2, v. 1, p. 17, pl. 2, fig. 17.

TYPE LOCALITY: British Isles.

AGE: Recent.

DESCRIPTION: Test elongate, somewhat pyriform, peripheral margin rounded; wall transparent, shiny except along opaque horseshoe shaped white band that runs along sides and aboral margin of test; pores larger in opaque band than elsewhere on test (Pl. 15, fig. 20); aboral margin sometimes spinose; aperture narrow, long, with parallel sides; entosolenian tube short, straight.

LENGTH: range: 130-280; mean: 180; coefficient of variation: 0.28.

DISTRIBUTION: see Figures 7 and 15.

Fissurina pulchella (Brady), Pl. 16, figs. 1-4.

ORIGINAL CITATION: *Lagena pulchella* Brady, 1870. Ann. Mag. Nat. Hist., ser. 4, v. 6, p. 294, pl. 12, fig. 1.

TYPE LOCALITY: Hebrides, Scotland.

AGE: Recent.

DESCRIPTION: Test subcircular to pyriform in outline, compressed, peripheral margin with narrow keel; wall opaque or translucent, with well developed rounded costae, most of which extend entire length of test although some are short and/or bifurcate; aperture fusiform, short, with thick lip.

LENGTH: range: 160-210.

DISTRIBUTION: see Figure 15.

Fissurina quadricostulata (Reuss), Pl. 16, figs. 5-7.

ORIGINAL CITATION: *Lagena quadricostulata* Reuss, 1870. K. Akad. Wiss. Wien, Math. Nat. Cl., v. 62, pt. 1, p. 469; Schlicht, 1870. Foram. Septarien-Thones. Pietzpuhl, pl. 6, figs. 25-30.

TYPE LOCALITY: Pietzpuhl, Germany.

AGE: Middle Oligocene.

DESCRIPTION: Test pyriform in outline, compressed, peripheral margin rounded, occasionally with spine on aboral end; wall thin, transparent to translucent, ornamented with two arcuate grooves parallel to border of each lateral face; aperture a long narrow slit.

LENGTH: range: 140-330; mean: 230; coefficient of variation: 0.17.

OBSERVATIONS: According to Reuss this species has 4 costae. However, examination under the SEM reveals that these "costae" are grooves, although they look like costae under ordinary light. Considering our photos and those of Kihle and Lofaldli (1973) we consider the description of Reuss to be erroneous. This is a widely distributed species both in time and space. Our collections have not revealed any specimens with 4 costae.

DISTRIBUTION: see Figures 7 and 15.

Fissurina semimarginata (Reuss), Pl. 16, figs. 8-10.

ORIGINAL CITATION: *Lagena marginata*, var. semimarginata Reuss, 1870. S.-B. Akad. Wiss. Wien, Math. Nat. Cl., v. 62, pt. 1, p. 468; Schlicht, 1870. Foram. Septarien-Thones. Pietzpuhl, p. 11, pl. 4, figs. 4-6.

TYPE LOCALITY: Pietzpuhl, Germany.

AGE: Middle Oligocene.

DESCRIPTION: Test pyriform in outline with produced and compressed apertural end, peripheral margin carinate; wall translucent, shiny, coarsely perforate except in central part of test; pores often elongate parallel to peripheral margin; aperture small, fusiform with thick lip; entosolenian tube long, straight.

LENGTH: range: 280-330.

DISTRIBUTION: see Figure 15.

FLORILUS Montfort, 1808

Test free, planispiral, involute, expanded, sometimes asymmetric, peripheral margin rounded to angulate; chambers increasing grandually in width and height; wall calcareous, perforate, granular, monolamellar; umbilici depressed, generally covered with granules that extend along sutures; aperture interiomarginal, central.

Florilus grateloupi (d'Orbigny), Pl. 16, figs. 11-14.

ORIGINAL CITATION: *Nonionina grateloupi* d'Orbigny, 1826. Ann. Scil Nat., sér. 1, v. 7, p. 294, no. 19; 1839. In: de la Sagra, Hist. Phys. Polit. Nat. Cuba, p. 46, pl. 6, figs. 6, 7.

TYPE LOCALITY: Cuba, Jamaica and Martinique.

AGE: Recent.

DESCRIPTION: Test elongate in outline, peripheral margin subacuate; chambers arcuate, expanding faster in breadth than height, 9-13 in final whorl; wall opaque to translucent, finely and densely perforate; sutures depressed, granular; umbilici small, depressed; aperture a well developed slit bordered by granules, lying at base of large elliptical apertural face.

DIAMETER: range: 280-800; mean: 500; coefficient of variation: 0.16.

DISTRIBUTION: see Figures 7 and 15.

Florilus pauperatus (Balkwill and Wright), Pl. 16, figs. 15-18.

ORIGINAL CITATION: *Nonionina pauperata* Balkwill and Wright, 1885. Roy. Irish Acad., Trans., v. 28, p. 353, pl. 13, figs. 25, 26.

TYPE LOCALITY: Lambay Island, Eire.

AGE: Recent.

DESCRIPTION: Test ovate in outline, lobate, peripheral margin angulate, keeled in early chambers; chambers slightly inflated, curved, 7-8 in final whorl, final chamber extended into umbilical region; wall translucent, smooth,

relatively large pores, widely spaced and concentrated in peripheral half of chambers; sutures limbate, narrow, somewhat depressed, aperture a semicircular slit partially filled by calcareous growth on keel, at base of triangular, flat, porous apertural face.

DIAMETER: range: 160-300; mean: 190; coefficient of variation: 0.21.

DISTRIBUTION: see Figures 7 and 15.

Florilus punctulatus (d'Orbigny), Pl. 16, figs. 19-21.

ORIGINAL CITATION: *Nonionina punctulata* d'Orbigny, 1839. Voy. Amér. Mérid., v. 5, pt. 5, p. 28, pl. 5, figs. 21, 22.

TYPE LOCALITY: Islas Malvinas.

AGE: Recent.

DESCRIPTION: Test ovate in outline, compressed, peripheral margin rounded, often asymmetrical in which case umbilical area is depressed; chambers low, wide, curved, 11-13 in final whorl, final ones strongly overhanging earlier ones in umbilical area; wall translucent, densely perforate; sutures arcuate, depressed; umbilicus depressed, slightly granulate; aperture an obscure slit at base of wide convex apertural face.

DIAMETER: range: 160-300.

DISTRIBUTION: see Figure 8.

GLABRATELLA Dorreen, 1948

Test free, inflated, trochospiral, sometimes streptospiral; chambers few, generally globose; wall calcareous, radial, hyaline, smooth or rugose, densely perforate; umbilical side smooth, polished and inscribed with radial striae; spiral side ornamented with calcareous granules; aperture umbilical, round.

Glabratella chasteri (Heron-Allen and Earland), Pl. 17, figs. 1-4.

ORIGINAL CITATION: *Discorbina minutissima* Chaster, 1892. Soc. Nat. Sci., v. 1, p. 65, pl. 1, figs. 15 = *Discorbina chasteri* Heron-Allen and Earland, 1913. Proc. Roy. Irish Acad., v. 31, pt. 64, p. 128, pl. 13, figs. 1-3.

TYPE LOCALITY: British Isles.

AGE: Recent.

DESCRIPTION: Test ovate in outline, compressed, peripheral margin rounded, lobate; chambers inflated to subglobular on spiral side, concave near umbilical region, 4-5 in final whorl, 1½-2 whorls; wall thin, transparent, shiny, finely perforate, on umbilical side convered with grooves radiating from umbilicus but failing to reach periphery; sutures depressed, well developed; aperture a large umbilical opening.

DIAMETER: range: 70-130.

DISTRIBUTION: see Figures 8 and 15.

GLOBULINA d'Orbigny, 1839

Test subspherical to ovate, trochospiral, circular to ovate in transverse section; early chambers quinqueloculine, later forming a triserial pattern with angles of 122-144°; wall calcareous, perforate; aperture terminal, radiate, cribrate, sometimes with entosolenian tube, sometimes fistulose.

Globulina australis d'Orbigny, Pl. 17, figs. 5-7.
ORIGINAL CITATION: *Globulina australis* d'Orbigny, 1839. Voy. Amér. Mérid., v. 5, pt. 5, p. 60, pl. 1, figs. 1-4.
TYPE LOCALITY: Bahía San Blas, Argentina.
AGE: Recent.
DESCRIPTION: Test gutiform in outline, somewhat compressed, ovate in transverse section; chambers elongate, gibbous, 2-3 in all; wall translucent, shiny, striate in aboral part of test and smooth to weakly striate in apertural area; sutures depressed; aperture radiate without entosolenian tube.
LENGTH: range: 240-540; mean: 380; coefficient of variation: 0.21.
OBSERVATIONS: There are but two poorly preserved specimens of this species in the Paris d'Orbigny collection.
DISTRIBUTION: see Figure 8.

Globulina caribaea (d'Orbigny), Pl. 17, figs. 8-11.
ORIGINAL CITATION: *Globulina caribaea* d'Orbigny, 1839. In: de la Sagra, Hist. Phys. Polit. Nat. Cuba, p. 130, pl. 2, figs. 7, 8.
TYPE LOCALITY: Not designated.
AGE: Recent.
DESCRIPTION: Test ovate in outline, somewhat compressed, ovate in transverse section, peripheral margin rounded; chambers large, somewhat inflated; wall translucent, finely perforate, ornamented with pustules; sutures slightly depressed; aperture radiate, small.
LENGTH: range: 380-700.
DISTRIBUTION: see Figure 8.

GUTTULINA d'Orbigny, 1839

Test ovate to elongate, asymmetric, trochospiral and quinqueloculine, generally triangular in transverse section; chambers often inflated, added in opposing series separated by 144°; wall calcareous, perforate; sutures generally depressed; aperture externally radiate, fisuriform, mamiform, or with rimmed circular orifices; internal apertural structure with tubiform canals, a single canal, a small tunnel located at apertural level, or an independent tube in the last chamber.

Guttulina lactea (Walker and Jacob), Pl. 17, figs. 12-14.
ORIGINAL CITATION: *Serpula lactea* Walker and Jacob, 1798. In: G. Adams, Essays Microsc., p. 634, pl. 14, fig. 4.

34

TYPE LOCALITY: Sandwich, Kent, England.
AGE: Recent.
DESCRIPTION: Test elongate in outline, peripheral margin rounded, somewhat compressed; chambers elongate, narrow; wall transparent to translucent, finely perforate, shiny, smooth; sutures well developed, slightly depressed; aperture radiate with or without entosolenian tube.
LENGTH: range: 170-340.
DISTRIBUTION: see Figures 8 and 15.

Guttulina plancii d'Orbigny, Pl. 17, figs. 15-17.
ORIGINAL CITATION: *Guttulina plancii* d'Orbigny, 1839. Voy. Amér. Mérid., v. 5, pt. 5, p. 60, pl. 1, fig. 5.
TYPE LOCALITY: Bahía San Blas, Argentina.
AGE: Recent.
DESCRIPTION: Test ovate to pyriform in outline, ovate in transverse section; peripheral margin rounded, somewhat lobate; chambers large, slightly inflated; wall translucent, smooth except in aboral region where covered with weak costae; sutures depressed; aperture radiate, sometimes with central pores.
LENGTH: range: 240-720.
DISTRIBUTION: see Figure 8.

Guttulina problema d'Orbigny, Pl. 17, figs. 18-20.
ORIGINAL CITATION: *Guttulina problema* d'Orbigny, 1826. Ann. Sci. Nat., sér. 1, v. 7, p. 266, no. 14, mod. 61.
TYPE LOCALITY: Castel Arquato, Italy.
AGE: Tertiary.
DESCRIPTION: Test pyriform in outline, peripheral margin rounded near base, subacute near aperture; chambers inflated, produced, variable; wall translucent, thick, smooth, finely perforate; sutures depressed; aperture radiate at margins, cribrate in center.
LENGTH: range: 230-670; mean: 410; coefficient of variation: 0.32.
DISTRIBUTION: see Figure 8.

GYPSINA Carter, 1877

Test attached, conical to discoidal; chambers circular, rectangular or polygonal, arranged in radial rows which alternate with those of earlier whorls; wall fibrous crystalline calcite, raised at surface forming an alveolar meshwork; aperture multiple in form of irregularly distributed pores.

Gypsina vesicularis (Parker and Jones), Pl. 18, figs. 1-3.
ORIGINAL CITATION: *Orbitolina vesicularis* Parker and Jones, 1860. Ann. Mag. Nat. Hist., ser. 3, v. 5, p. 31, Brady, 1884. Challenger Exp., Repts., Zool., v. 9, p. 718, pl. 101, figs. 9-12.
TYPE LOCALITY: Pacific Ocean.
AGE: Recent.

DESCRIPTION: Test outline conical, peripheral margin sometimes subacute; chambers numerous, small, polygonal, raised, arranged in radiating, encrusting, hemispherical mass, alternating with one another in successive layers; wall white, shiny, translucent, rough; apertures formed by irregularly distributed small openings in wall of some chambers.

DIAMETER: range: 330-360.

DISTRIBUTION: see Figure 8.

HANZAWAIA Asano, 1944

Test free, trochospiral, periphery angulate, spiral side planar with cameral extension on lower margin of chambers partially covering chambers of previous whorl, umbilical side convex, involute; wall calcareous, granular, densely perforate except on apertural face; sutures limbate, strongly curved; aperture interiomarginal, arcuate, may extend onto spiral side under central flap of last chamber; supplementary apertures below chamber extensions.

Hanzawaia boueana (d'Orbigny), Pl. 18, figs. 4-8.

ORIGINAL CITATION: *Truncatulina boueana* d'Orbigny, 1846. Foram. Foss. Bass. Tert. Vienne, p. 169, pl. 9, figs. 24-26.

TYPE LOCALITY: Nussdorf, Vienna.

AGE: Tertiary.

DESCRIPTION: Test ovate to subcircular in outline, lobate, peripheral margin carinate, variable, spiral side planar or weakly convex, umbilical side convex with slightly depressed umbilicus; chambers curved, 7-9 in final whorl, internal border on spiral side produced into arched flap which partially covers earlier whorls; wall translucent, finely perforate except on apertural face; sutures limbate, strongly curved, raised or flush in early chambers, depressed in later chambers; aperture a peripheral arc bordered by thin lip, extending along spiral suture for 3-4 chambers; supplementary apertures on spiral side beneath cameral flaps.

DIAMETER: range: 170-600.

DISTRIBUTION: see Figure 8. Small poorly developed and isolated forms of this species occur on the inner shelf south of the distribution shown on Figure 8 and also in rare amounts in the Malvin Current.

HERONALLENIA Chapman and Parr, 1931

Test free, trochospiral, plano-convex, compressed, peripheral margin carinate; chambers concave and few on spiral side, inflated and bearing radial grooves on umbilical side; umbilical area depressed; wall calcareous, radial, perforate; sutures raised on spiral side, depressed on umbilical side; aperture ovate, large, umbilical.

Heronallenia kempii (Heron-Allen and Earland), Pl. 18, figs. 9-13.

ORIGINAL CITATION: *Discorbis kempii* Heron-Allen and Earland, 1929. Roy. Micr. Soc., Jour., ser. 3, v. 49, p. 332, pl. 4, figs. 40-48.

TYPE LOCALITY: Banco Burdwood, Islas Malvinas and Estrecho de Magallanes.

AGE: Recent.

DESCRIPTION: Test ovate in outline, lobate, peripheral margin rounded to carinate, compressed; chambers on spiral side ornamented with irregularly arranged and raised costae which ramify from raised sutures, on umbilical side, inflated, with inner margin concave, bearing radiating intersutural furrows which may or may not reach peripheral margin, 6-8 in final whorl; wall translucent, glassy, finely perforate; aperture eccentric, at base of ovate depression lying beneath internal margin of chambers, with short triangular tooth.

DIAMETER: range: 360-920.

OBSERVATIONS: *Heronallenia* requires additional study and redescription. There is no doubt that the species described above is the same described by Heron-Allen and Earland from the same area. However, with the SEM it is possible to note transverse sutures on the umbilical side corresponding to septa that partition the chambers into two parts, an internal and an external. This characteristic has not yet been described in the literature.

DISTRIBUTION: see Figures 8 and 15.

HOEGLUNDINA Brotzen, 1948

Test free, lenticular, trochospiral; chambers with well developed internal partition extending from posterior wall of chamber; umbilicus closed; wall calcareous, radial, aragonite, perforate; primary aperture ovate, interiomarginal, extraumbilical; secondary apertures nearly peripheral, extending breadth of chamber, earlier ones open or secondarly closed by addition of shell material.

Hoeglundina elegans (d'Orbigny), Pl. 18, figs. 14-17.

ORIGINAL CITATION: *Rotalia (Turbinulina) elegans* d'Orbigny, 1826. Ann. Sci. Nat., sér. 1, v. 7, p. 276, no. 54 = *Pulvinulina elegans* (d'Orbigny), Brady, 1884. Challenger Exp., Repts., Zool., v. 9, p. 699, pl. 105, figs. 4-6.

TYPE LOCALITY: Coroncina, Italy.

AGE: Not designated.

DESCRIPTION: Test subcircular in outline, biconvex to plano-convex; peripheral margin acute and carinate, 2-2½ whorls; chambers triangular on umbilical side, quadrate on spiral side, 6-7 in final whorl; wall transparent, shiny, finely perforate, with irregular blotches of white shell material; sutures limbate, distinct, curved on spiral side, radial and straight on umbilical side; primary aperture small, absent in some specimens; supplementary apertures an arcuate slit with flanges on both sides but smaller on internal side.

DIAMETER: range: 530-1250.

DISTRIBUTION: see Figure 15.

HOPKINSINA Howe and Wallace, 1932

Test elongate, initial portion triserial, later biserial and slightly torted; wall calcareous, perforate, radial, smooth or ornamented with pustules, striae or costae; aperture terminal, oval or circular, usually with short neck, lip and tooth plate.

Hopkinsina pacifica Cushman, Pl. 18, figs. 18-20.

ORIGINAL CITATION: *Hopkinsina pacifica* Cushman, 1933. Cushman Lab. Foram. Res., Contr., v. 9, pt. 4, p. 86, pl. 8, fig. 16.

TYPE LOCALITY: Tonga Islands.

AGE: Recent.

DESCRIPTION: Test ovate in transverse section, peripheral margin rounded, lobate; chambers inflated, broad, high; wall finely pustulate; sutures depressed; apertural neck askew on final chamber, with a well developed lip.

LENGTH: range: 110-240; mean: 190; coefficient of variation: 0.16.

DISTRIBUTION: see Figure 8.

LAGENA Walker and Jacob, 1798

Test unilocular, spherical, ovoid or fusiform; wall calcareous, perforate, smooth or ornamented with costae, pustules, reticulation, etc.; aperture simple, rounded, with elongate neck and non-radiate, phialine lip.

Lagena aspera Reuss, Pl. 19, figs. 1-3.

ORIGINAL CITATION: *Lagena aspera* Reuss, 1861. S.-B. Akad. Wiss. Wien, v. 44, p. 305, pl. 11, fig. 5.

TYPE LOCALITY: Maastricht, Netherlands.

AGE: Cretaceous.

DESCRIPTION: Test globular; wall thin, with irregularly distributed coarse and truncated pustules which are sometimes united (Pl. 19, fig. 3); apertural neck short with uniform diameter; aperture small.

LENGTH: range: 140-210.

DISTRIBUTION: see Figures 8 and 15.

Lagena caudata (d'Orbigny), Pl. 19, figs. 4-7.

ORIGINAL CITATION: *Oolina caudata* d'Orbigny, 1839. Voy. Amér. Mérid., v. 5, pt. 5, p. 19, pl. 5, fig. 6.

TYPE LOCALITY: Islas Malvinas.

AGE: Recent.

DESCRIPTION: Test elongate, somewhat fusiform, inflated at aboral end; wall thin, transparent to translucent, shiny,

completely striate or only from the base halfway up test; short basal spine; aperture a small circular hole on long tapering, fine, striated neck with well developed phialine lip.

LENGTH: range: 210-610.

DISTRIBUTION: see Figures 8 and 15.

Lagena clavata (d'Orbigny), Pl. 19, figs. 8, 9.

ORIGINAL CITATION: *Oolina clavata* d'Orbigny, 1846. Foram. Foss. Bass. Tert. Vienne, p. 24, pl. 1, figs. 2, 3.

TYPE LOCALITY: Baden, south of Vienna.

AGE: Tertiary.

DESCRIPTION: Test elongate, fusiform with long delicate neck and well developed spinose extension of aboral part of test, circular in transverse section; wall translucent, thin, smooth; aperture rounded, small, with phialine lip.

LENGTH: range: 440-1080.

DISTRIBUTION: see Figures 8 and 15.

Lagena digitale Heron-Allen and Earland, Pl. 19, figs. 10-12.

ORIGINAL CITATION: *Lagena digitale* Heron-Allen and Earland, 1932. Discovery Repts., v. 4, p. 371, pl. 10, figs. 28-30.

TYPE LOCALITY: Estrecho de Magallanes and Tierra del Fuego.

AGE: Recent.

DESCRIPTION: Test ovate in outline, circular in transverse section; wall opaque, with strong honeycomb ornamentation; aperture small at end of cone shaped neck of variable length (Pl. 19, fig. 12).

LENGTH: range: 230-410.

DISTRIBUTION: see Figures 8 and 15.

Lagena distoma Parker and Jones, sensu lato, Pl. 19, figs. 13-17.

ORIGINAL CITATION: *Lagena laevis* (Montagu), var. striata Parker and Jones, 1860. Ann. Mag. Nat. Hist., ser. 2, v. 19, p. 278, pl. 11, figs. 24 = *Lagena distoma* Parker and Jones, ms., 1864. Brady, Linn. Soc., Trans., v. 24, p. 467, pl. 48, fig. 6.

TYPE LOCALITY: Coast of Norway, Shetland Islands, Northcumberland coast, England.

AGE: Recent.

DESCRIPTION: Test fusiform, elongate, terminating in long, delicate, gradually tapering necks at both ends, central portion inflated or with parallel sides; wall translucent, thin, shiny, with fine longitudinal striae; apertures at both ends, but lacking phialine lip at one end (Pl. 19, fig. 17).

OBSERVATIONS: This species occurs in two formae.

Lagena distoma Parker and Jones, forma typica, Pl. 19, figs. 13, 14.

CHARACTERISTIC TRAITS: This forma is characterized by its parallel sides.

LENGTH: range: 570-1070.

DISTRIBUTION: see Figure 15.

Lagena distoma Parker and Jones, forma turgida, Pl. 19, figs. 15-17.

ORIGINAL CITATION: *Lagena distoma* Parker and Jones, forma turgida Boltovskoy, 1961. Neotropica, v. 7, no. 24, p. 76, fig. 7.

LOCALITY: Brazilian shelf, 23-24°S.

AGE: Recent.

CHARACTERISTIC TRAITS: This forma is distinguished from forma typica by its more inflated and rounded test.

LENGTH: range: 640-1420.

DISTRIBUTION: see Figure 15.

Lagena gracilis Williamson, Pl. 19, figs. 18, 19.

ORIGINAL CITATION: *Lagena gracilis* Williamson, 1848. Ann. Mag. Nat. Hist., ser. 1, v. 1, p. 13, pl. 1, fig. 5.

TYPE LOCALITY: Boston, England.

AGE: Recent.

DESCRIPTION: Test elongate, tapering gradually at apertural end, rapidly at aboral end, sometimes with aboral spine; wall transparent, thin, with 14-18 striae some of which extend along the long apertural neck; aperture without phialine lip.

LENGTH: range: 430-530.

OBSERVATIONS: This species is quite variable in its outline, the number of striae, and the nature of the aboral end. The specimen described and illustrated by Williamson was fusiform at both ends, but in a later work Williamson (1858) described this species with a rounded aboral end. Specimens from the southwest Atlantic are not as fusiform as those originally described by Williamson, nor are they as rounded at the aboral end as those described by him or by Murray (1971).

DISTRIBUTION: see Figure 16.

Lagena hispidula Cushman, Pl. 19, figs. 20-22.

ORIGINAL CITATION: *Lagena hispidula* Cushman, 1913. U.S. Nat. Mus., Bull. 71, pt. 3, p. 14, pl. 5, figs. 2, 3.

TYPE LOCALITY: Nero Station 1264, between Yokohama and Guam.

AGE: Recent.

DESCRIPTION: Test flask shaped, with aboral end rounded; apertural neck thin and rather long; wall covered with small irregularly shaped pustules (Pl. 19, fig. 22), finely perforate; aperture rounded, small.

LENGTH: range: 330-470.

DISTRIBUTION: see Figures 8 and 16.

Lagena cf. *L. interrupta* Williamson, Pl. 20, figs. 1-3.

ORIGINAL CITATION: *Lagena striata*, var. α interrupta Williamson, 1848. Ann. Mag. Nat. Hist., ser. 2, v. 1, p. 14, pl. 1, fig. 7.

TYPE LOCALITY: British Isles.

AGE: Recent.

DESCRIPTION: Test club-shaped, terminating in long tapering apertural neck; wall translucent, covered from one extreme to other with quadrate longitudinal costae which alternate with shorter, finer, more rounded costae; aperture small with well developed phialine lip.

LENGTH: range: 270-510.

OBSERVATIONS: *L. interrupta* has a test which is more globose and relatively shorter than those specimens from the southwest Atlantic. Consequently we consider this form to be *L.* cf. *L. interrupta*.

DISTRIBUTION: see Figures 8 and 16.

Lagena laevis (Montagu), sensu lato, Pl. 20, figs. 4-10.

ORIGINAL CITATION: *Vermiculum laeve* Montagu, 1803. Testac. Brit., p. 524. *Lagena laevis* (Montagu), Williamson, 1848. Ann. Mag. Nat. Hist., ser. 1, v. 1, p. 12, pl. 1, figs. 1, 2.

TYPE LOCALITY: Sandwich, Kent, England.

AGE: Recent.

DESCRIPTION: Test outline lagenaform, usually asymmetrical, apertural end tapering gradually and ending in thick walled neck occasionally bearing fine irregular costae at the extreme end, aboral end smooth or bearing short costae; wall smooth, glassy; aperture rounded with weak phialine lip.

OBSERVATIONS: There are two distinct formae of this species in the area.

Lagena laevis (Montagu), forma typica, Pl. 20, figs. 4-6.

CHARACTERISTIC TRAITS: This forma has a smooth aboral end.

LENGTH: range: 240-410; mean: 310; coefficient of variation: 0.16.

DISTRIBUTION: see Figures 8 and 16. This species also occurs in the brackish waters of Lagoa dos Patos, Arroio Chuí, and Río Quequén.

Lagena laevis (Montagu), forma tenuis, Pl. 20, figs. 7-10.

ORIGINAL CITATION: *Ovulina tenuis* Bornemann, 1855. Dt. Geol. Ges., v. 7, no. 2, p. 317, pl. 12, fig. 3.

TYPE LOCALITY: Seasalter, Kent, England.

AGE: Recent.

CHARACTERISTIC TRAITS: This forma is distinguished from forma typica by its short costae on the aboral extremity.

LENGTH: range: 210-530; mean: 350; coefficient of variation: 0.26.

DISTRIBUTION: see Figure 8. This species also occurs in the brackish waters of Mar Chiquita.

Lagena striata (d'Orbigny), Pl. 20, figs. 11-14.

ORIGINAL CITATION: *Oolina striata* d'Orbigny, 1839. Voy. Amér. Mérid., v. 5, pt. 5, p. 21, pl. 5, fig. 12.

TYPE LOCALITY: Islas Malvinas.

AGE: Recent.

DESCRIPTION: Test semicircular in outline, rarely ovate, with long gradually tapering neck; wall thin, with fine costae that cover entire test, intercalated with shorter costae, aboral extremity with labyrinthic pattern; aperture small, rounded and located at the end of a neck whose sides are ornamented in polygonal patterns.
LENGTH: range: 330-510; mean: 440; coefficient of variation: 0.14.
OBSERVATIONS: Several typical specimens of this species can be found in the original material of d'Orbigny in Paris.
DISTRIBUTION: see Figures 8 and 16.

Lagena sulcata (Walker and Jacob), forma *lyellii*, Pl. 20, figs. 15-17.
ORIGINAL CITATION: *Amphorina lyellii* Seguenza, 1862. Desc. Foram. Monotal. Messina, p. 52, pl. 1, fig. 40.
TYPE LOCALITY: Rometta, Messina, Sicily, Italy.
AGE: Tertiary.
DESCRIPTION: Test onion shaped, apertural end with neck 1/3 length of test, aboral end with thick short spine; wall translucent, with variable number of weak costae which lie principally on aboral half of test and occasionally reaching neck; aperture small on striate or smooth neck, with lobate phialine lip.
LENGTH: range: 210-650.
DISTRIBUTION: see Figures 9 and 16.

LOXOSTOMUM Ehrenberg, 1854

Test free, elongate, compressed, initially biserial, later uniserial; chambers numerous; wall calcareous, granular, perforate, smooth or ornamented; sutures arcuate, limbate; aperture terminal, oval or slit like, with lip, beginning above basal suture.

Loxostomum albatrossi (Cushman), Pl. 20, figs. 18-22.
ORIGINAL CITATION: *Bolivina albatrossi* Cushman, 1922. U.S. Nat. Mus., Bull. 104, pt. 3, p. 31, pl. 6, fig. 4.
TYPE LOCALITY: California coast, USA.
AGE: Recent.
DESCRIPTION: Test ovate in outline, peripheral margin rounded; chambers elongate, oblique, expanding very slowly as added, 6-7 pairs; wall irregularly alveolate and crenulate in early portion of test, smooth in last few chambers; pores located in irregularly distributed alveolar pits (Pl. 20, fig. 22); sutures oblique, hidden by strong ornamentation in early chambers, depressed in later chambers; aperture terminal, ovate, short, with weak rim.
LENGTH: range: 190-330; mean: 250; coefficient of variation: 0.16.
OBSERVATIONS: All the specimens studied from the southwest Atlantic area had a terminal aperture whose base did not coincide with the basal suture; hence the designation as *Loxostomum*.
DISTRIBUTION: see Figure 9.

38

MASSILINA Schlumberger, 1893

Test ovate in outline, commonly compressed, initially quinqueloculine, later biloculine; chambers partially embracing; wall calcareous, porcellaneous, imperforate; aperture terminal with tooth.

Massilina secans (d'Orbigny), Pl. 21, figs. 1-4.
ORIGINAL CITATION: *Quinqueloculina secans* d'Orbigny, 1826. Ann. Sci. Nat., sér. 1, v. 7, p. 303, no. 43, mod. 96.
TYPE LOCALITY: Adriatic Sea and Mediterranean Sea.
AGE: Recent.
DESCRIPTION: Test broadly ovate in outline, peripheral margin carinate or rounded; wall smooth or with weak striae; sutures depressed; aperture ovate, with simple or bifid tooth which is sometimes surrounded by a weak neck.
LENGTH: range: 260-700.
DISTRIBUTION: see Figure 9.

MELONIS Montfort, 1808

Test free, planispiral, symmetric, involute, peripheral margin rounded, biumbilicate; umbilici deep, surrounded by imperforate calcite; wall calcareous, perforate, granular, bilamellar; sutures flush or slightly depressed, straight or slightly curved; apertural face smooth, imperforate; aperture interiomarginal, extending on both sides of periphery toward umbilicus.

Melonis affine (Reuss), Pl. 21, figs. 5-7.
ORIGINAL CITATION: *Nonionina affinis* Reuss, 1851. Z. Dt. Geol. Ges., v. 3, p. 72, pl. 5, fig. 32.
TYPE LOCALITY: Hermsdorf, Germany.
AGE: Eocene.
DESCRIPTION: Test ovate, compressed, peripheral margin broadly rounded; 9-11 chambers in final whorl; wall translucent, densely perforate with relatively large pores; sutures limbate, expanding toward umbilicus, smooth, curved, slightly depressed in final few chambers; aperture lunate with narrow border, located at base of flat, imperforate apertural face.
DIAMETER: range: 230-530; mean: 390; coefficient of variation: 0.26.
DISTRIBUTION: see Figures 9 and 16.

MILIOLINELLA Wiesner, 1931

Test quinqueloculine or triloculine; wall calcareous, porcellaneous, imperforate; sutures distinct; aperture terminal, covered by valvular flap.

Miliolinella lutea (d'Orbigny), Pl. 21, figs. 8-10.
ORIGINAL CITATION: *Triloculina lutea* d'Orbigny, 1839. Voy. Amér. Mérid., v. 5, pt. 5, p. 70, pl. 9, figs. 6-8.
TYPE LOCALITY: Islas Malvinas.

AGE: Recent.

DESCRIPTION: Test ovate in outline, peripheral margin rounded, ovate in transverse section; chambers inflated, expanding toward apertural end and strongly embracing the penultimate chamber; wall smooth, thin, transparent; sutures distinct; aperture an arcuate slit, with well developed valvular tooth.

LENGTH: range: 270-600; mean: 380; coefficient of variation: 0.21.

DISTRIBUTION: see Figures 9 and 16.

Miliolinella subrotunda (Montagu), Pl. 21, figs. 11-14.

ORIGINAL CITATION: *Vermiculum subrotundum* Montagu, 1803. Testac. Brit., p. 512 = *Miliolina subrotunda* Brady, 1884. Challenger Exp., Repts., Zool., v. 9, p. 168, pl. 5, figs. 10, 11.

TYPE LOCALITY: British Isles.

AGE: Recent.

DESCRIPTION: Test ovate to suborbicular in outline, somewhat compressed in transverse section, peripheral margin rounded; wall shiny, generally smooth, occasionally with very weak transverse and longitudinal striations; sutures depressed; aperture semicircular with small valvular tooth.

LENGTH: range: 200-550; mean: 350; coefficient of variation: 0.29.

OBSERVATIONS: The population of *M. subrotunda* is composed of both quinqueloculine and triloculine specimens, with the latter being more numerous among adult forms.

DISTRIBUTION: see Figures 9 and 16. This species also occurs in the brackish waters of the Río de la Plata and Río Quequén.

MORULAEPLECTA Höglund, 1947

Test initially streptospiral, later biserial; wall agglutinated; aperture an interiomarginal slit.

Morulaeplecta bulbosa Höglund, Pl. 21, figs. 15, 16.

ORIGINAL CITATION: *Morulaeplecta bulbosa* Höglund, 1947. Zool. Bidr. Uppsala, v. 26, p. 165, pl. 12, fig. 2, textfig. 142.

TYPE LOCALITY: Gullmar Fjord.

AGE: Recent.

DESCRIPTION: Test elongate, tapering toward aboral end, initial end somewhat flat and bulky, coiled in a closed streptospiral of 5-7 chambers; chambers somewhat inflated, 6-7 pairs in biserial portion; wall composed of irregular, rugose grains; sutures somewhat depressed, straight, perpendicular to growth axis; aperture small, ovate.

LENGTH: range: 200-410; mean: 300; coefficient of variation: 0.26.

DISTRIBUTION: see Figures 9 and 16.

MYCHOSTOMINA Berthelin, 1881

Test free, conical or concavo-convex with dorsal side concave; on dorsal side the subspherical proloculus is followed by a simple tubular second chamber which passes through several volutions; on the ventral side the second chamber crosses the periphery and coils toward the umbilicus for $1\frac{1}{2}$-2 whorls and has a narrow peripheral keel; wall composed of 1-3 crystals of calcite, hyaline, variable thickness, rough with growth lines, coarsely to finely perforate; pores vary in size, number, and distribution on different specimens; growth lines distinct; aperture terminal, umbilical, partially covered by dorsal growth from external side of tube.

Mychostomina revertens (Rhumbler), Pl. 21, figs. 17-20.

ORIGINAL CITATION: *Spirillina vivipara* Ehrenberg, var. revertens Rhumbler, 1906. Zool. Jb., Abt. Syst. Geogr. Biol., v. 24, p. 32, pl. 2, figs. 8-10.

TYPE LOCALITY: Hawaiian Islands and Marshall Islands.

AGE: Recent.

DESCRIPTION: Test discoidal, peripheral margin subacute, convexo-concave; wall finely perforate with pores outlining spire; sutures slightly depressed; aperture central in shape of crescent.

DIAMETER: range: 160-280.

DISTRIBUTION: see Figure 9.

NONION Montfort, 1808

Test planispiral, involute to asymmetric, compressed, biumbonate, peripheral margin rounded, sometimes lobate; chambers numerous; wall calcareous, perforate, granular; umbilici tuberculate, canaliculate; aperture interiomarginal, equatorial slit, sometimes multiple openings.

Nonion depressulus (Walker and Jacob), Pl. 22, figs. 1-5.

ORIGINAL CITATION: *Nautilus depressulus* Walker and Jacob, 1798. In: Adams, Essays Microsc., p. 641, pl. 14, fig. 33.

TYPE LOCALITY: Reculver, Kent, England.

AGE: Recent.

DESCRIPTION: Test subcircular in outline, flattened, peripheral margin rounded, slightly lobate; chambers narrow, arcuate, 10-14 in final whorl; wall thin, translucent, shiny, finely and densely perforate; sutures curved, excavated, bordered with pustules (Pl. 22, fig. 5) which continue into wide umbilicus; apertural face convex, perforate; aperture composed of small holes surrounded by pustules.

DIAMETER: range: 280-720; mean: 470; coefficient of variation: 0.21.

DISTRIBUTION: see Figure 9.

NONIONELLA Cushman, 1926

Test free, weakly trochospiral, asymmetric, partially evolute, somewhat compressed, peripheral margin rounded to acute; chambers numerous on umbilical side, final chamber overhanging umbilicus; wall calcareous, perforate, granular; aperture interiomarginal, arcuate, extending toward umbilicus.

Nonionella atlantica Cushman, Pl. 22, figs. 6-9.
ORIGINAL CITATION: *Nonionella atlantica* Cushman, 1947. Cushman Lab. Foram. Res., Contr., v. 23, pt. 4, p. 90, pl. 20, figs. 4, 5.
TYPE LOCALITY: Canaveral, Florida, USA.
AGE: Recent.
DESCRIPTION: Test ovate in outline, peripheral margin subacute; chambers slightly inflated, curved, 10-12 in final whorl; spiral side partially evolute; umbilical side involute; umbilicus covered with granules and papillae which may unite to form crests; wall smooth, white, finely and densely perforate; sutures depressed, curved, granulate near umbilicus; apertural face wide, convex; aperture bordered by one or two series of granules.
DIAMETER: range: 300-580.
DISTRIBUTION: see Figure 9. This species also occurs in the brackish waters of Lagoa dos Patos.

Nonionella auris (d'Orbigny), Pl. 22, figs. 10-12.
ORIGINAL CITATION: *Valvulina auris* d'Orbigny, 1839. Voy. Amér. Mérid., v. 5, pt. 5, p. 47, pl. 2, figs. 15-17.
TYPE LOCALITY: South American coast.
AGE: Recent.
DESCRIPTION: Test large, robust, ovate in outline, slightly lobate, peripheral margin subcarinate; chambers inflated, slightly curved, 9-11 in final whorl, final with lobate, digitate extension that covers umbilicus on involute side; spiral side partially evolute, flattened; wall smooth, white, finely and densely perforate; sutures very depressed, especially near umbilicus; apertural face broad, triangular, convex; aperture distinct.
DIAMETER: range: 210-780; mean: 520; coefficient of variation: 0.27.
OBSERVATIONS: Within the populations of this species there are specimens which, compared to *N. auris*, are smaller, with fewer chambers, and less well developed. These specimens were described from the same area by Heron-Allen and Earland as *N. iridea*.
DISTRIBUTION: see Figures 9 and 16.

Nonionella chiliensis Cushman and Kellett, Pl. 22, figs. 13-16.
ORIGINAL CITATION: *Nonionella chiliensis* Cushman and Kellett, 1929, U.S. Nat. Mus., Proc., v. 75, art. 25, p. 6, pl. 2, fig. 4.
TYPE LOCALITY: Corral, Chile.

AGE: Recent.
DESCRIPTION: Test ovate to circular in outline, peripheral margin subacute, spiral side slightly convex with 2-2½ volutions, umbilical side with depressed, slightly granulate umbilicus; chambers low, broad, 10-12 in final whorl; wall smooth, translucent to opaque, finely and densely perforate; sutures limbate, depressed, curved; apertural face triangular with finely granulate border surrounding aperture.
DIAMETER: range: 170-400.
DISTRIBUTION: see Figure 9.

Nonionella pulchella Hada, Pl. 22, figs. 17-20.
ORIGINAL CITATION: *Nonionella pulchella* Hada, 1931. Tohoku Univ. Sci. Rep., ser. 4 (Biol.), v. 6, p. 120, textfig. 79.
TYPE LOCALITY: Mutsu Bay, Japan.
AGE: Recent.
DESCRIPTION: Test ovate in outline, slightly lobate, compressed, peripheral margin subacute to rounded, spiral side partially evolute with large, depressed, circular umbilicus; chambers slightly inflated, curved, 8-11 in final whorl, the last of which has a lobate, digitate extension over umbilicus on involute side; wall smooth, translucent, densely and finely perforate; sutures depressed, curved; apertural face triangular, narrow, convex.
DIAMETER: range: 200-380; mean: 250; coefficient of variation: 0.16.
OBSERVATIONS: This species is closely related to *N. auris* and may be a junior synonym of it. However, in the area of study we are able to distinguish clearly between the two on the following bases: *N. auris* is larger, more robust, wider, and has deeper sutures.
DISTRIBUTION: see Figure 9.

NOTOROTALIA Finlay, 1939

Test free, trochospiral, biconvex to plano-convex, compressed; umbilical area generally covered by chamber extensions underlain by spiral canal system; sutures generally with double row of sutural pores which originate from a subsutural canal; wall calcareous, perforate, ornamented with ridges running parallel to periphery; aperture with one or two rows of subcircular openings near base of flat apertural face; apertural face covered by radiating irregular ridges.

Notorotalia clathrata (Brady), Pl. 23, figs. 1-3.
ORIGINAL CITATION: *Rotalia clathrata* Brady, 1884. Challenger Exp., Repts., Zool., v. 9, p. 709, pl. 107, figs. 8, 9.
TYPE LOCALITY: Between Australia and New Zealand.
AGE: Recent.
DESCRIPTION: Test subcircular in outline, peripheral margin subacute to carinate, 3 volutions, the last of which

with 7-9 low arcuate chambers; wall translucent, ornamented with crests which generally lie transverse to sutures; intercrestal areas covered with small granules; sutural crests on umbilical side lie in depressed groove; aperture covered by crests and granules.

DIAMETER: range: 170-550; mean: 330; coefficient of variation: 0.30.

DISTRIBUTION: see Figure 9.

OOLINA d'Orbigny, 1839

Test free, unilocular, ovate or globular, circular in transverse section; wall hyaline, perforate, smooth or ornamented with striations, costae or reticulations; aboral region may bear a spine; aperture rounded, simple, sometimes radiate, with entosolenian tube.

Oolina acuticosta (Reuss), Pl. 23, figs. 4-6.
ORIGINAL CITATION: *Lagena acuticosta* Reuss, 1862. K. Akad. Wiss., Wien, v. 44, p. 305, pl. 1, fig. 4.
TYPE LOCALITY: Maastricht, Netherlands.
AGE: Cretaceous.
DESCRIPTION: Test pyriform, aboral part rounded, apertural end elongate; wall ornamented with 10-12 thick, acuminate costae; aperture small, lying at top of broad collar.
LENGTH: range: 240-340.
DISTRIBUTION: see Figure 16.

Oolina borealis Loeblich and Tappan, Pl. 23, figs. 7, 8.
ORIGINAL CITATION: *Entosolenia costata* Williamson, 1858. Foram. Gr. Brit., Ray Soc., p. 9, pl. 1, figs. 18 [emend. Loeblich and Tappan, 1954, Wash. Acad. Sci., Jour., v. 44, no. 12, p. 384, *Oolina borealis*].
TYPE LOCALITY: Isle of Skye, Scotland.
AGE: Recent.
DESCRIPTION: Test ovate in outline, aboral end rounded; wall translucent with thick costae which are quadrangular and somewhat concave in section; aperture small, round, at top of collar which may bear weak costae.
LENGTH: range: 230-310.
DISTRIBUTION: see Figures 9 and 16. This species also occurs in the brackish waters of the Lagoa dos Patos.

Oolina caudigera (Wiesner), Pl. 23, figs. 9-12.
ORIGINAL CITATION: *Lagena (Entosolenia) globosa*, var. caudigera Wiesner, 1931. Dt. Südpol. Exp., v. 20 (Zool., v. 12), p. 119, pl. 18, fig. 214.
TYPE LOCALITY: Antarctica.
AGE: Recent.
DESCRIPTION: Test irregularly ovate in outline, sometimes laterally compressed, often with asymmetrically placed aboral spine; wall thin, translucent, smooth; aperture small, surrounded with pustules; entosolenian tube straight and usually long.

LENGTH: range: 170-410.
DISTRIBUTION: see Figures 9 and 16.

Oolina globosa (Montagu), Pl. 23, figs. 13, 14.
ORIGINAL CITATION: *Serpula (Lagena) laevis globosa* Walker and Boys, 1784. Minute Shells, p. 3, pl. 1, fig. 8 = *Vermiculum globosum* Montagu, 1803. Testac. Brit., p. 523.
TYPE LOCALITY: Sandwich, Kent, England.
AGE: Recent.
DESCRIPTION: Test gutiform in outline; wall smooth, opaque, white; aperture simple, large.
LENGTH: range: 210-400.
DISTRIBUTION: see Figures 9 and 16.

Oolina hexagona (Williamson), Pl. 23, figs. 15-17.
ORIGINAL CITATION: *Entosolenia squamosa*, var. hexagona Williamson, 1848. Ann. Mag. Nat. Hist., ser. 2, v. 1, p. 20, pl. 2, fig. 23.
TYPE LOCALITY: British Isles.
AGE: Recent.
DESCRIPTION: Test ovate in outline; wall opaque, ornamented with hexagonal reticulate pattern which sometimes becomes irregular at extreme ends; aperture small, circular.
LENGTH: range: 210-270.
DISTRIBUTION: see Figures 9 and 16.

Oolina lineata (Williamson), Pl. 23, figs. 18-21.
ORIGINAL CITATION: *Entosolenia lineata* Williamson, 1848. Ann. Mag. Nat. Hist., ser. 2, v. 1, p. 18, pl. 2, fig. 18.
TYPE LOCALITY: British Isles.
AGE: Recent.
DESCRIPTION: Test elongate or gutiform in outline, aboral end rounded, occasionally with short spine, apertural end truncate; wall translucent or opaque, ornamented with numerous fine costae which begin at aboral end and may or may not reach aperture; apertural wall thickened; aperture radiate.
LENGTH: range: 160-240.
DISTRIBUTION: see Figures 9 and 16.

Oolina melo d'Orbigny, Pl. 24, figs. 1-5.
ORIGINAL CITATION: *Oolina melo* d'Orbigny, 1839. Voy. Amér. Mérid., v. 5, pt. 5, p. 20, pl. 5, fig. 9.
TYPE LOCALITY: Islas Malvinas.
AGE: Recent.
DESCRIPTION: Test ovate in outline, rounded in aboral part, acuminate in apertural end; wall translucent or opaque, ornamentation reticulate with longitudinal costae which are connected by transverse, concave downward costae; aperture small; entosolenian tube short.
LENGTH: range: 170-260; mean: 210; coefficient of variation: 0.10.
OBSERVATIONS: *O. melo* is a highly variable species, par-

ticularly with respect to its ornamentation and to the presence of a short apertural collar. Within the populations in the study area are specimens which are similar to *O. hexagona* and very similar to *O. squamosa* from which some forms of *O. melo* are difficult to distinguish.

DISTRIBUTION: see Figures 9 and 16. This species also occurs in the brackish waters of the Río de la Plata and Río Quequén.

Oolina squamosa (Montagu), Pl. 24, figs. 6-8.
ORIGINAL CITATION: *Vermiculum squamosum* Montagu, 1803. Testac. Brit., p. 526, pl. 14, fig. 2.
TYPE LOCALITY: Seasalter, Kent, England.
AGE: Recent.
DESCRIPTION: Test ovate in outline, aboral end rounded, apertural end somewhat acuminate; wall translucent or opaque with scale-like ornamentation consisting of longitudinal costae which are connected by short slightly arcuate costae; aperture small, with short neck.
LENGTH: range: 210-270.
OBSERVATIONS: Variants of this species with a very short or no neck grade into the concept of *O. melo*.
DISTRIBUTION: see Figures 10 and 16.

Oolina vilardeboana d'Orbigny, Pl. 24, figs. 9-11.
ORIGINAL CITATION: *Oolina vilardeboana* d'Orbigny, 1839. Voy. Amér. Mérid., v. 5, pt. 5, p. 19, pl. 5, figs. 4, 5.
TYPE LOCALITY: Islas Malvinas.
AGE: Recent.
DESCRIPTION: Test pyriform in outline, aboral part rounded, apertural end tapered with an ornamental collar; wall translucent with numerous (up to 25), high, occasionally bifurcating, longitudinal costae which are rectangular in section; aperture small, circular.
LENGTH: range: 280-530.
DISTRIBUTION: see Figures 10 and 16.

ORTHOMORPHINA Stainforth, 1952

Test rectilinear, uniserial; wall calcareous, perforate, smooth or costate; aperture terminal, rounded, with a weak collar or inverted ring.

Orthomorphina calomorpha (Reuss), Pl. 24, figs. 12, 13.
ORIGINAL CITATION: *Nodosaria calomorpha* Reuss, 1866. K. Akad. Wiss. Wien, Math., Nat. Cl., v. 25, pt. 1, p. 129, pl. 1, figs. 15-19.
TYPE LOCALITY: Pietzpuhl, Germany.
AGE: Middle Oligocene.
DESCRIPTION: Test elongate, rectilinear or slightly arcuate, composed of 2-3 chambers, the first of which is the smallest and most rounded, the second of which is the most elongate; wall translucent, very thin, finely perforate; sutures depressed, perpendicular to growth axis; aperture rounded

or elongate, with well developed ring, without collar.
LENGTH: range: 210-430.
DISTRIBUTION: see Figures 10 and 17.

Orthomorphina filiformis (d'Orbigny)?, Pl. 24, figs. 14-16.
ORIGINAL CITATION: *Nodosaria filiformis* Soldani, 1789-1798. *Testaceogr. Zoophytogr.*, v. 2, p. 35, pl. 10, fig. e. D'Orbigny, 1826. Ann. Sci. Nat., sér. 1, v. 7, p. 253, no. 14.
TYPE LOCALITY: Sienna, Tuscany, Italy.
AGE: Not designated.
DESCRIPTION: Test elongate, arcuate; chambers of equal diameter throughout length of test, 4-5, inflated, elliptical; wall smooth, translucent, shiny, finely perforate; sutures distinct, depressed, perpendicular to growth axis; aperture small.
LENGTH: range: 260-780.
OBSERVATIONS: We are not certain of the specific identity of this species because the specimens from the area of study lack a collar and have fewer chambers than that of the original illustration. Moreover, several investigators, including Brady, have placed this species in *Dentalina*. Unfortunately, the location of the primary types is unknown and topotypes are unavailable, so that it is impossible to establish the true character of this foraminiferal species.
DISTRIBUTION: see Figure 10.

PATELLINA Williamson, 1858

Test free, trochospiral, spiral side elevated and evolute, umbilical side planar and involute, microspheric forms with an elliptical proloculus followed by an undivided tubular chamber which is wound in a spiral of 3 volutions, megalospheric forms with a small proloculus followed by two wide and low chambers per whorl; initial chambers divided by transverse and incomplete secondary septa which are intercalated with a set of short tertiary septa; wall calcareous, monocrystalline, finely perforate, with line of pits on spiral side; aperture umbilical, arcuate, lying beneath the exterior margin of a flap.

Patellina corrugata Williamson, Pl. 24, figs. 17-20.
ORIGINAL CITATION: *Patellina corrugata* Williamson, 1858. Foram. Gr. Brit., Ray Soc., p. 46, pl. 3, figs. 86-89.
TYPE LOCALITY: British Isles.
AGE: Recent.
DESCRIPTION: Test circular in outline, conical, peripheral margin angulate and carinate, spiral side convex, umbilical side planar to concave; chambers on spiral side lunate and occupying more than half a whorl, with line of distinct pores parallel to sutures; wall thin, translucent, with growth lines; sutures limbate on spiral side; aperture elongate, located on internal edge of last chamber.

LENGTH: range: 140-570; mean: 260; coefficient of variation: 0.42.

DISTRIBUTION: see Figures 10 and 17.

PLANORBULINA d'Orbigny, 1826

Test attached, initial portion trochospiral, later planispiral; chambers added in cycles; wall calcareous, perforate, radial; aperture interiomarginal, at both ends of each chamber, with lip.

Planorbulina mediterranensis d'Orbigny, Pl. 25, figs. 1-3.
ORIGINAL CITATION: *Planorbulina mediterranensis* d'Orbigny, 1826. Ann. Sci. Nat., sér. 1, v. 7, p. 280, no. 2, pl. 14, figs. 4-6, mod. 79.
TYPE LOCALITY: Not designated (probably Mediterranean Sea).
AGE: Recent.
DESCRIPTION: Test subcircular in outline, lobate, peripheral margin rounded, spiral side convex, umbilical side planar or concave, 2-6 whorls; chambers on spiral side globose, quadrangular, arcuate to irregular in shape, on umbilical side flat, embracing; wall white to yellow white, thin, translucent to opaque, coarsely perforate; sutures depressed; apertures in final whorl lunate with protruding well developed lip.
DIAMETER: range: 990-1180.
DISTRIBUTION: see Figure 10.

POROEPONIDES Cushman, 1944

Test free, trochospiral, plano-convex to biconvex, peripheral margin angulate and carinate, umbilical area excavated; chambers numerous; wall calcareous, radial, perforate; sutures oblique, curved on spiral side, radial on umbilical side; aperture an interiomarginal arch extending from umbilicus to periphery, with narrow rim and umbilical flap projecting from central part of internal chamber margin; secondary apertures areal, scattered on apertural face.

Poroeponides lateralis (Terquem), pl. 25, figs. 4-7.
ORIGINAL CITATION: *Rosalina lateralis* Terquem, 1878. Mém. Soc. Géol., sér. 3, pt. 1, p. 25, pl. 2, fig. 11.
TYPE LOCALITY: Isle of Rhodes.
AGE: Late Pliocene.
DESCRIPTION: Test subcircular in outline, peripheral margin acute and carinate; chambers triangular in early part, becoming semi-lunate, 6-8 in final whorl; wall smooth, shiny, white, finely perforate except in region of secondary apertures; sutures depressed on umbilical side, distinct, flush or raised on spiral side; aperture a narrow, long slit.

DIAMETER: range: 400-1180; mean: 750; coefficient of variation: 0.31.
DISTRIBUTION: see Figure 10. This species also occurs in the mixohaline waters of Lagoa dos Patos.

PSAMMOSPHAERA Schulze, 1875

Test free, generally spherical, unilocular; wall agglutinated, well cemented; aperture indefinite.

Psammosphaera fusca Schulze, Pl. 25, fig. 8.
ORIGINAL CITATION: *Psammosphaera fusca* Schulze, 1875. 2^d Jahresb. Komm. Untersuch. Dt. Meere Kiel, p. 113, pl. 2, fig. 8.
TYPE LOCALITY: Coast of Norway.
AGE: Recent.
DESCRIPTION: Test spherical or ovate; wall composed of few large or many small grains that protrude sharply from test, variable composition; cement and internal wall reddish white, barely visible between grains.
DIAMETER: range: 540-990.
DISTRIBUTION: see Figure 17.

PULLENIA Parker and Jones, 1862

Test free, globular or lenticular, planispiral, involute; final whorl with 3-6 chambers; wall calcareous, granular, perforate; sutures radial; aperture interiomarginal, narrow, elongate.

Pullenia bulloides (d'Orbigny), Pl. 25, figs. 9-11.
ORIGINAL CITATION: *Nonionina bulloides* d'Orbigny, 1826. Ann. Sci. Nat., sér. 1, v. 7, p. 127 [293], no. 2. 1846. Foram. Foss. Bass. Tert. Vienne, p. 107, pl. 5, figs. 9, 10.
TYPE LOCALITY: Nussdorf, Vienna.
AGE: Tertiary.
DESCRIPTION: Test subspherical, slightly compressed, peripheral margin rounded; 4-5 chambers in final whorl; wall smooth, white, finely perforate; sutures distinct, flush or slightly depressed; apertural face low; aperture reaching umbilical region.
DIAMETER: range: 280-440.
DISTRIBUTION: see Figure 17.

Pullenia subcarinata subcarinata (d'Orbigny), Pl. 25, figs. 12, 13.
ORIGINAL CITATION: *Nonionina subcarinata* d'Orbigny, 1839. Voy. Amér. Mérid., v. 5, pt. 5, p. 28, pl. 5, figs. 23, 24.
TYPE LOCALITY: Islas Malvinas.
AGE: Recent.
DESCRIPTION: Test subcircular to ovate in outline, bilaterally symmetrical, convex, peripheral margin subacute; 6 (rarely 5) chambers in final whorl; wall smooth, shiny,

finely perforate, white; sutures slightly depressed, weakly curved; aperture a slit with thick lip.

DIAMETER: range: 280-740; mean: 500; coefficient of variation: 0.22.

OBSERVATIONS: There are six excellently preserved specimens of this species in the d'Orbigny collection in Paris.

This subspecies has a smoother peripheral outline, more chambers and is wider in apertural view than *P. subcarinata quinqueloba*.

DISTRIBUTION: see Figures 10 and 17.

Pullenia subcarinata quinqueloba (Reuss), Pl. 25, figs. 14, 15.

ORIGINAL CITATION: *Nonionina quinqueloba* Reuss, 1851. Z. Dt. Geol. Ges., v. 3, p. 47, pl. 5, fig. 31.

TYPE LOCALITY: Hermsdorf, Berlin.

AGE: Eocene.

DESCRIPTION: Test subcircular in outline, slightly lobate, compressed, peripheral margin subacute; 4-5 chambers in final whorl; wall smooth, shiny, finely perforate, white; sutures depressed, slightly curved; aperture a long slit with bordering lip.

DIAMETER: range: 280-440.

OBSERVATIONS: This subspecies is distinguished from *P. subcarinata subcarinata* by having a more compressed test, a more lobate periphery and fewer chambers in the final whorl.

DISTRIBUTION: see Figure 10.

PYRGO Defrance, 1824

Test free, ovate, biloculine; chambers inflated, embracing; microspheric forms initially quinqueloculine, later triloculine, finally biloculine; megalospheric forms biloculine only; wall calcareous, imperforate, porcellaneous; aperture terminal, near union of two chambers; tooth usually bifid.

Pyrgo elongata (d'Orbigny), Pl. 25, figs. 16, 17.

ORIGINAL CITATION: *Biloculina elongata* d'Orbigny, 1826. Ann. Sci. Nat., sér. 1, no. 7, p. 298, no. 4 = *Miliola (Biloculina) elongata* d'Orbigny, Parker and Jones, 1865. Roy. Soc. London, Philos. Trans., v. 155, p. 409, pl. 17, figs. 88, 90, 91.

TYPE LOCALITY: Pauliac, Gironde, France.

AGE: Not designated.

DESCRIPTION: Test elongate, tapering toward apertural end; chambers rounded at peripheral margins; sutures somewhat depressed; wall smooth, shiny; aperture horseshoe shaped, tooth expanding at its distal end.

LENGTH: range: 270-550.

OBSERVATIONS: This species is closely related to *P. ringens* and the two may represent environmental variants of each other.

44

DISTRIBUTION: see Figure 10.

Pyrgo nasuta Cushman, Pl. 25, figs. 18-21.

ORIGINAL CITATION: *Pyrgo nasuta* Cushman, 1935. Smithson. Inst., Misc. Coll., v. 91, no. 21, p. 7, pl. 3, figs. 1-4.

TYPE LOCALITY: North of Puerto Rico.

AGE: Recent.

DESCRIPTION: Test subcircular in outline, peripheral outline of final chamber terminating in right angle with respect to penultimate chamber; chambers inflated; wall smooth, shiny, white; sutures distinct; aperture an arcuate slit on compressed neck, with broad tooth projecting into aperture.

LENGTH: range: 330-890; mean: 560; coefficient of variation: 0.23.

DISTRIBUTION: see Figure 10.

Pyrgo peruviana d'Orbigny, Pl. 26, figs. 1-3.

ORIGINAL CITATION: *Biloculina peruviana* d'Orbigny, 1839. Voy. Amér. Mérid., v. 5, pt. 5, p. 68, pl. 9, figs. 1-3.

TYPE LOCALITY: Payta, Peru.

AGE: Recent.

DESCRIPTION: Test ovate to circular in outline, compressed, subcircular to elliptical in transverse section; chambers rounded; wall smooth, shiny, white; sutures depressed; aperture a curved slit with short tooth on its upper surface and a low, short, expanding or bifurcating tooth on its lower surface.

LENGTH: range: 510-870.

DISTRIBUTION: see Figures 10 and 17.

Pyrgo quadrata (Heron-Allen and Earland), Pl. 26, figs. 4-6.

ORIGINAL CITATION: *Biloculina elongata* d'Orbigny, var. quadrata Heron-Allen and Earland, 1930. Roy. Micr. Soc., Jour., ser. 3, v. 50, p. 50, pl. 2, figs. 1-4.

TYPE LOCALITY: Devonshire, England.

AGE: Recent.

DESCRIPTION: Test square to rectangular in outline, compressed; chambers flattened; wall smooth, shiny; sutures distinct; aperture a large slit with wide spatulate tooth whose ends are somewhat lobate; apertural border with weak central salient on upper part.

LENGTH: range: 430-1780.

DISTRIBUTION: see Figure 10.

Pyrgo ringens (Lamarck), Pl. 26, figs. 7-9.

ORIGINAL CITATION: *Miliolites (ringens) subglobosa* Lamarck, 1804. Ann. Mus. Nat. Hist., pt. 5, p. 351, v. 9, pl. 17, fig. 1.

TYPE LOCALITY: Grignon, Paris, France.

AGE: Eocene.

DESCRIPTION: Test elongate, peripheral margin rounded; chambers bulky and pyriform; wall smooth, shiny, white;

sutures depressed; aperture subcircular with bifid tooth.
LENGTH: range: 340-990; mean: 580; coefficient of variation: 0.31.
DISTRIBUTION: see Figure 10.

Pyrgo subsphaerica (d'Orbigny), Pl. 26, figs. 10-13.
ORIGINAL CITATION: *Biloculina subsphaerica* d'Orbigny, 1839. In: de la Sagra, Hist. Phys. Polit. Natur. Cuba, p. 162, pl. 8, figs. 25-27.
TYPE LOCALITY: Cuba and Jamaica.
AGE: Recent.
DESCRIPTION: Test circular in outline, ovate in transverse section; chambers globose; wall smooth, shiny; sutures depressed, occasionally sinuous; aperture large, ovate with high upper border and a bifurcating tooth.
LENGTH: range: 150-440.
DISTRIBUTION: see Figure 10.

QUINQUELOCULINA d'Orbigny, 1826

Test free, ovate; chambers enrolled about a transverse axis in quinqueloculine fashion, maintaining an angle of 144° with the preceeding chamber, 5 visible at any time, each 72° from adjacent chamber; wall generally calcareous, porcellaneous, imperforate, sometimes agglutinate; aperture terminal, rounded or elongate, with simple or bifid tooth.

Quinqueloculina angulata (Williamson), Pl. 26, figs. 14-17.
ORIGINAL CITATION: *Milliolina bicornis* (Walker and Jacob), var. angulata Williamson, 1858. Foram. Gr. Brit., Ray Soc., p. 88, pl. 7, fig. 196.
TYPE LOCALITY: British Isles.
AGE: Recent.
DESCRIPTION: Test elongate; chamber margins angulate; wall with weak irregular striations; apertural end of final chamber slightly expanded; aperture narrow with long, thin, simple tooth.
LENGTH: range: 170-850; mean: 550; coefficient of variation: 0.29.
OBSERVATIONS: In the northern part of the study area the chambers are more angular and protruded than in the southern areas where the borders are more rounded.
DISTRIBUTION: see Figure 10.

Quinqueloculina arctica Cushman, Pl. 26, figs. 18-20.
ORIGINAL CITATION: *Quinqueloculina arctica* Cushman, 1933. Smithson. Inst., Misc. Coll., v. 89, no. 9, p. 2, pl. 1, fig. 3.
TYPE LOCALITY: Clavering Island, Greenland.
AGE: Recent.
DESCRIPTION: Test robust; chambers angulate; wall thick, somewhat rough; sutures distinct; aperture rounded with tooth which expands toward its distal end.

LENGTH: range: 470-1260.
DISTRIBUTION: see Figure 10.

Quinqueloculina atlantica Boltovskoy, Pl. 27, figs. 1-3.
ORIGINAL CITATION: *Quinqueloculina atlantica* Boltovskoy, 1957. Inst. Nac. Inv. Cienc. Nat., Rev., Geol., v. 6, no. 1, p. 25, pl. 5, figs. 2-6.
TYPE LOCALITY: 35°27'5"S; 55°04'00"W.
AGE: Recent.
DESCRIPTION: Test triangular in transverse section; apertural end of final chamber expanded, producing arched apertural border; wall smooth; sutures distinct; aperture narrow, elongate, with long thin tooth.
LENGTH: range: 370-1290; mean: 590; coefficient of variation: 0.36.
DISTRIBUTION: see Figure 10.

Quinqueloculina brodermanni Seiglie, Pl. 27, figs. 4-7.
ORIGINAL CITATION: *Quinqueloculina brodermanni* Seiglie, 1965. Cushman Found. Foram. Res., Contr., v. 16, pt. 2, p. 71, pl. 8, figs. 3, 4.
TYPE LOCALITY: Península Araya, Venezuela.
AGE: Recent.
DESCRIPTION: Test subcircular in outline, triangular in transverse section; chambers raised and weakly angulate; wall agglutinate; sutures obscure; aperture horseshoe-shaped with short simple tooth.
LENGTH: range: 300-810.
DISTRIBUTION: see Figure 10. This species also occurs in the mixohaline waters of the Río de la Plata.

Quinqueloculina frigida Parker, Pl. 27, figs. 8-12.
ORIGINAL CITATION: *Quinqueloculina frigida* Parker, 1952. Harvard Coll., Mus. Comp. Zool., Bull., v. 106, p. 406, pl. 3, fig. 20.
TYPE LOCALITY: 43°04'N; 70°33.6'W.
AGE: Recent.
DESCRIPTION: Test broadly triangular in transverse section, peripheral margin rounded, side with three chambers visible concave; wall agglutinated with well sorted, flush, dark gray grains and black, irregularly distributed grains; sutures distinct; aperture rounded, relatively small with short, simple tooth.
LENGTH: range: 310-680; mean: 430; coefficient of variation: 0.19.
OBSERVATIONS: This species is variable with respect to its size, wall color, and the degree of convexity of the side with three chambers. The most characteristic trait of this species is the presence of distinct black grains distributed randomly over the test surface.
DISTRIBUTION: see Figure 10.

Quinqueloculina gregaria Andreae, Pl. 27, figs. 13-16.
ORIGINAL CITATION: *Quinqueloculina gregaria* Andreae, 1884. Geol. Spezialk-Karte Elsass Loth., Abh., v. 2, pt. 3,

p. 186, pl. 12, figs. 10-12.

TYPE LOCALITY: Elsass.

AGE: Late Oligocene.

DESCRIPTION: Test compressed, peripheral margin rounded to weakly angulate; chambers flat or somewhat produced; wall thick, white, with deep longitudinal striae; sutures distinct; aperture ovate with long bifid tooth with or without a short neck.

LENGTH: range: 600-1290.

OBSERVATIONS: This species was identified by Heron-Allen and Earland (1932) as *Q. inca*, but we disagree with this determination in that *Q. inca* as described and drawn by d'Orbigny has very angulate margins. Unfortunately the original material of d'Orbigny cannot be located.

DISTRIBUTION: see Figure 11.

Quinqueloculina horrida Cushman, Pl. 27, figs. 17-20.

ORIGINAL CITATION: *Quinqueloculina horrida* Cushman, 1947. Cushman Lab. Foram. Res., Contr., v. 23, p. 88, pl. 19, fig. 1.

TYPE LOCALITY: Charleston, South Carolina, USA.

AGE: Recent.

DESCRIPTION: Test outline acute at aboral and apertural end, triangular in transverse section, peripheral margin rounded; chambers indistinct; wall agglutinate, rough, composed of angular to subangular quartz grains; sutures indistinct; aperture on short cylindrical neck, small, with short tooth.

LENGTH: range: 190-510.

DISTRIBUTION: see Figure 11.

Quinqueloculina intricata Terquem, Pl. 28, figs. 1-8.

ORIGINAL CITATION: *Quinqueloculina intricata* Terquem, 1878. Mém. Soc. Géol., sér. 3, pt. 1, p. 73, pl. 8, figs. 16-21.

TYPE LOCALITY: Isle of Rhodes.

AGE: Pliocene.

DESCRIPTION: Test irregular in outline, elongate, peripheral margin angulate; final chambers curved, produced, quadrate and carinate, early chambers more rounded; wall white, thick, costate; sutures depressed; aperture circular with short simple tooth located on long neck.

LENGTH: range: 260-1160; mean: 550; coefficient of variation: 0.42.

OBSERVATIONS: Within the populations of this species are many specimens which we interpret as juvenile forms following the viewpoint of Terquem (Pl. 28, figs. 3, 6).

DISTRIBUTION: see Figure 11.

Quinqueloculina lamarckiana d'Orbigny, Pl. 28, figs. 9-12.

ORIGINAL CITATION: *Quinqueloculina lamarckiana* d'Orbigny, 1839. In: de la Sagra, Hist. Phys. Polit. Natur. Cuba, p. 189, pl. 11, figs. 14, 15.

TYPE LOCALITY: Cuba and Jamaica.

AGE: Recent.

DESCRIPTION: Test rounded in aboral part, truncated at apertural end, triangular in transverse section; chambers angulate and carinate; wall white, shiny; sutures almost flush; aperture ovate with long simple tooth.

LENGTH: range: 280-1070; mean: 630; coefficient of variation: 0.33.

OBSERVATIONS: The majority of the specimens of this species in the Río de la Plata have a triloculine arrangement similar to *Triloculina tricarinata*. The strongly carinate and angulate forms (Pl. 28, figs. 9-11) are typical of low latitudes whereas higher latitude forms loose their angularity and exhibit more rounded margins (Pl. 28, fig. 12).

DISTRIBUTION: see Figure 11.

Quinqueloculina milletti (Wiesner), Pl. 28, figs. 13-17.

ORIGINAL CITATION: *Miliolina bosciana* (d'Orbigny), 1898. Millett (pars), Jour. Roy. Micr. Soc., p. 267, pl. 6, fig. 1 = *Miliolina milletti* Wiesner, 1912. Arch. Protistenk., v. 25, p. 220, 237.

TYPE LOCALITY: Isle of Ischia, Italy.

AGE: Recent.

DESCRIPTION: Test with peripheral margin rounded, rarely angulate, early visible chambers oblique or erect; wall thin, shiny, sometimes finely striate; sutures distinct; aperture large, ovate with simple small tooth or open.

LENGTH: range: 160-440; mean: 250; coefficient of variation: 0.24.

OBSERVATIONS: This species is quite variable morphologically and exhibits several traits apart from those of the typical specimen described above:

a) presence of longitudinal striae and costae (in Río Quequén)

b) presence of a short neck (Uruguayan coast)

c) an angulate transverse section (Uruguayan coast)

d) parallel margins (in Río Quequén)

DISTRIBUTION: see Figure 11. This species is also found in the brackish waters of the Río de la Plata, Mar Chiquita and Río Quequén.

Quinqueloculina patagonica d'Orbigny, Pl. 28, figs. 18-21.

ORIGINAL CITATION: *Quinqueloculina patagonica* d'Orbigny, 1839. Voy. Amér. Mérid., v. 5, pt. 5, p. 84, pl. 4, figs. 14-16.

TYPE LOCALITY: South of the mouth of the Río Negro, Argentina.

AGE: Recent.

DESCRIPTION: Test elongate, triangular in transverse section, peripheral margin rounded; chambers of equal width throughout length; wall smooth, white, porcellaneous; sutures depressed; aperture ovate with simple, long, bifid tooth.

LENGTH: range: 270-1020; mean: 450; coefficient of variation: 0.36.

OBSERVATIONS: This species is variable in its outline (degree of parallelism of sides), the development of a neck, and the transverse section (compressed or triangulate).

DISTRIBUTION: see Figure 11.

Quinqueloculina polygona d'Orbigny, Pl. 29, figs. 1-6.

ORIGINAL CITATION: *Quinqueloculina polygona* d'Orbigny, 1839. In: de la Sagra, Hist. Phys. Polit. Natur. Cuba, p. 198, pl. 12, figs. 21-23.

TYPE LOCALITY: Cuba and Jamaica.

AGE: Recent.

DESCRIPTION: Test elongate, aboral end rounded, apertural end truncate, angulate to rounded in transverse section, side with three chambers concave, side with four chambers convex; chambers quadrate, often bicarinate; wall finely and irregularly striate and grooved (Pl. 29, fig. 6); sutures depressed.

LENGTH: range: 150-420.

OBSERVATIONS: We interpret this species as *Q. polygona* with some reservation because according to the description and drawings of d'Orbigny it is more angulate than the specimens we have observed in the area. Our specimens show both angulate (Pl. 29, figs. 1, 3; but not as angulate as d'Orbigny) and rounded (Pl. 29, figs. 2, 4) forms as well as those intermediate between these extremes.

DISTRIBUTION: see Figure 11.

Quinqueloculina seminulum (Linné), Pl. 29, figs. 7-13.

ORIGINAL CITATION: *Serpula seminulum* Linné, 1767. Syst. Nat., ed. 12, p. 1264, no. 791.

TYPE LOCALITY: Adriatic Sea.

AGE: Recent.

DESCRIPTION: Test with peripheral margin rounded; chambers of uniform diameter throughout length; wall smooth, shiny; sutures depressed; aperture semicircular with rim and thin bifid tooth.

LENGTH: range: 380-1420; mean: 740; coefficient of variation: 0.39.

OBSERVATIONS: *Q. seminula* is a cosmopolitan species. In addition to its occurrence in the inner shelf, some specimens occur in the outer shelf. The species is morphologically variable and its environmental variants have been described under a variety of names. In the study area these include: *Q. magellanica*, *Q. isabellei* and *Q. araucana* (see Heron-Allen and Earland, 1932; Boltovskoy, 1954a).

DISTRIBUTION: see Figures 11 and 17. This species also occurs in the brackish waters of the Río de la Plata, Arroio Chuí and Lagoa dos Patos.

Quinqueloculina stalkeri Loeblich and Tappan, Pl. 29, figs. 14-16.

ORIGINAL CITATION: *Quinqueloculina stalkeri* Loeblich and Tappan, 1953. Smithson. Misc. Coll., v. 121, no. 7, p. 40, pl. 5, figs. 5-9.

TYPE LOCALITY: Godthaab Island, Greenland.

AGE: Recent.

DESCRIPTION: Test with rounded peripheral margin; chambers curved, inflated and uniform diameter throughout length; wall irregularly covered with finely agglutinated grains; sutures depressed; neck short; aperture ovate to rounded with rim and small simple short tooth.

LENGTH: range: 140-320.

DISTRIBUTION: see Figure 11.

RECURVOIDES Earland, 1934

Test free, subglobular, planispiral in initial portion, streptospiral in adult at right angles to initial coil; wall agglutinated; aperture small, areal, with lip.

Recurvoides contortus Earland, Pl. 29, figs. 17-20.

ORIGINAL CITATION: *Recurvoides contortus* Earland, 1933. Discovery Repts., v. 7, p. 91, pl. 10, figs. 7-19.

TYPE LOCALITY: South Georgia Islands and Antarctic.

AGE: Recent.

DESCRIPTION: Test subcircular to ovate in outline, peripheral margin rounded, faintly lobate; chambers rectangular or trapezoidal, 8-9 in final whorl, arranged in streptospiral coil; umbilical zone depressed on involute side, planar on evolute side; wall finely agglutinate, smooth; sutures slightly depressed, straight; aperture ovate, areal, with distinct lip.

LENGTH: range: 510-970.

DISTRIBUTION: see Figure 17.

REMANEICA Rhumbler, 1938

Test attached, trochospiral, compressed; chambers numerous, low; wall agglutinated, yellow-white, with well developed rugae on umbilical side; sutures curved; aperture unclear, apparently on umbilical side margin of last chamber, as an umbilical slit.

Remaneica helgolandica Rhumbler, Pl. 30, figs. 1-4.

ORIGINAL CITATION: *Remaneica helgolandica* Rhumbler, 1938. Kiel Meeresf., v. 2, no. 2, p. 195, text-figs. 38-42.

TYPE LOCALITY: Northern Netherlands.

AGE: Recent.

DESCRIPTION: Test subcircular in outline, highly compressed, peripheral margin acute, spiral side convex and evolute, umbilical side concave and partially involute, 3 volutions; chambers increasing very slowly in size as added, 9-14 in final whorl, several folds along septal suture on umbilical side terminating in triangular lobe which overhangs the depressed umbilicus; wall thin, smooth, shiny, finely agglutinate; sutures depressed, lobulate; aperture obscure, very small, ovate, with lip, interiomarginal, extra-umbilical.

LENGTH: range: 160-380.

OBSERVATION: Many specimens from the southwest Atlantic area posses more chambers than described by Rhumbler.

DISTRIBUTION: see Figure 11.

REOPHAX Montfort, 1808

Test free, elongate, straight or arcuate, circular or ovate in transverse section, uniserial; wall agglutinated, little cement and grains of various materials, with rough texture: sutures perpendicular to growth axis: aperture terminal, circular to fissuriform, with tubular neck.

Reophax curtus Cushman, Pl. 30, figs. 5-7.

ORIGINAL CITATION: *Reophax curtus* Cushman, 1920. U.S. Geol. Surv., Prof. Paper 128-B, p. 8, pl. 2, figs. 2, 3.

TYPE LOCALITY: 46°48'30''N, 52°43'W.

AGE: Recent.

DESCRIPTION: Test irregularly conical, straight or slightly curved, circular in transverse section; chambers expanding rapidly in size as added, 3-5; wall composed of agglutinated quadrangular grains; sutures depressed, sometimes indistinguishable due to agglutinated material; aperture small, terminal, irregularly shaped.

LENGTH: range: 1010-1990; mean: 1560; coefficient of variation: 0.15.

DISTRIBUTION: see Figure 17.

Reophax scorpiurus Montfort, Pl. 30, figs. 8-10.

ORIGINAL CITATION: *Reophax scorpiurus* Montfort, 1808. Conchyol. Syst., v. 1, p. 331, text-fig. 130.

TYPE LOCALITY: Beaches of the Adriatic Sea.

AGE: Recent.

DESCRIPTION: Test rectilinear or curved, steplike in outline, chambers expanding considerably in size, 3-6, circular in section, inflated; wall rough, composed of poorly sorted grains giving chambers an irregular outline, brownish yellow; sutures depressed; aperture irregularly ovate.

LENGTH: range: 850-1790.

DISTRIBUTION: see Figure 17.

ROBULUS Montfort, 1808

Test free, planispiral, bilaterally symmetrical, lenticular, involute, biumbonate with raised central umbilici, peripheral margin subangular, carinate, or keeled; chambers numerous, wide; wall calcareous, radial, smooth or ornamented, finely perforate; sutures radial, curved; aperture radiate, at peripheral angle with elongate slit in medial line of apertural face, often surrounded by lateral extensions of the wall of the final chamber.

Robulus limbosus (Reuss), sensu lato, Pl. 30, figs. 11-14.

ORIGINAL CITATION: *Robulina limbosa* Reuss, 1863. S.-B. Akad. Wiss. Wien, v. 48, p. 55, pl. 6, fig. 69.

TYPE LOCALITY: Not designated.

AGE: Not designated.

DESCRIPTION: Test suborbicular or slightly elongate in outline, peripheral margin acute or keeled, variable biconvexity; chambers arcuate, 7-9 in final whorl; umbilical area convex, covered with glassy calcite through which early whorls are visible; wall smooth, transparent to translucent; sutures limbate, wide; apertural face concave or planar with sharp or vague limits; aperture radiate with narrow vertical slit which may have lip.

DIAMETER: range: 310-850.

DISTRIBUTION: see Figure 11.

Robulus orbicularis (d'Orbigny), Pl. 30, figs. 15-17.

ORIGINAL CITATION: *Robulina orbicularis* d'Orbigny, 1826. Ann. Sci. Nat., sér. 1, v. 7, p. 288, pl. 15, figs. 8, 9.

TYPE LOCALITY: Vicinity of Sienna, Tuscany, Italy.

AGE: Not designated.

DESCRIPTION: Test subcircular in outline, markedly involute. peripheral margin with narrow keel; chambers strongly curved, narrow, 6-8 in final whorl; umbilici wide, convex, transparent; wall smooth, glassy; sutures limbate, sharply curved to rear; aperture produced, with short radial bars.

DIAMETER: range: 400-710.

DISTRIBUTION: see Figures 11 and 17.

Robulus rotulatus (Lamarck), sensu lato, Pl. 30, figs. 18-20; Pl. 31, figs. 1, 2.

ORIGINAL CITATION: *Lenticulites (rotulata)* Lamarck, 1804. Ann. Mus. Nat., v. 5, p. 188, pl. 62, fig. 11.

TYPE LOCALITY: Meudon, Paris.

AGE: Late Cretaceous.

DESCRIPTION: Test circular to subcircular in outline, peripheral margin acute to keeled; chambers low, triangular, arcuate, 6-10 in final whorl; wall translucent to opaque, glassy in umbonal and sutural areas; umbilical area convex with a knob of variable size; sutures limbate, thick, arcuate; apertural face concave or planar; aperture large, radiate, with raised vertical fissure.

OBSERVATIONS: There are two formae of this species in the southwest Atlantic area.

Robulus rotulatus (Lamarck), forma typica, Pl. 30, figs. 18-20.

CHARACTERISTIC TRAITS: This forma is distinguished by its acute periphery and absence of a keel.

DIAMETER: range: 300-950.

DISTRIBUTION: see Figures 11 and 17.

Robulus rotulatus (Lamarck), forma cultrata, Pl. 31, figs. 1, 2.

ORIGINAL CITATION: *Robulus cultrata* Montfort, 1808. Conchyol. System., v. 1, p. 214, 54-e genre.

TYPE LOCALITY: Coroncine, Tuscany, Italy.

AGE: Not designated.

CHARACTERISTIC TRAITS: This forma is distinguished from forma typica by the presence of a keel.

DIAMETER: range: 540-1160.

OBSERVATIONS: This forma is distinguished from *R. limbosus* by being completely involute and from *R. orbicularis* by having less strongly curved sutures.

DISTRIBUTION: see Figures 11 and 17.

ROLSHAUSENIA Bermúdez, 1952

Test free, trochospiral, unequally biconvex, umbilical side more convex than spiral; chambers numerous; umbilicus covered with calcareous material; wall calcareous, hyaline, finely perforate; primary aperture complicated, an elongate interiomarginal slit connected to previous chamber; secondary apertures separated from primary by septal plate, opening into umbilicus beneath a cameral flap; additional supplementary apertures, semilunate on anterior margin of umbilical flap of each chamber.

Rolshausenia rolshauseni (Cushman and Bermúdez), Pl. 31, figs. 3-5.

ORIGINAL CITATION: *Rotalia rolshauseni* Cushman and Bermúdez, 1946. Cushman Lab. Foram. Res., Contr., v. 22, pl. 119, pl. 19, figs. 11-13.

TYPE LOCALITY: Texas coast, USA.

AGE: Recent.

DESCRIPTION: Test subcircular in outline, peripheral margin rounded, strongly lobate, spiral side with 2½ volutions; chambers, weakly inflated, 5-7 in final whorl; sutures open, depressed, with thickened margins, curved backwards slightly; wall opaque, densely perforate; primary aperture an umbilical-extraumbilical slit.

DIAMETER: range: 280-510.

DISTRIBUTION: see Figure 11.

SACCAMMINA M. Sars, 1869

Test free, subspherical, sometimes pyriform, unilocular, circular in transverse section; wall agglutinated with large grains, strongly cemented with white to yellow material; aperture irregularly circular.

Saccammina atlantica (Cushman), Pl. 31, figs. 6-8.

ORIGINAL CITATION: *Proteonina atlantica* Cushman, 1944. Cushman Lab. Foram. Res., Sp. Publ. 12, p. 5, pl. 1, fig. 4.

TYPE LOCALITY: Vineyard Sound, Massachusetts, USA.

AGE: Recent.

DESCRIPTION: Test pyriform; wall completely covered with large grains, usually quartz, giving test a blocky outline, whitish; aperture irregularly circular.

DIAMETER: range: 370-510.

DISTRIBUTION: see Figures 11 and 17.

SIGMOILINA Schlumberger, 1887

Test free, ovate, in microspheric form early chambers added in triloculine fashion, angle increases gradually from 120° to 180° in adult form to give test sigmoidal curve in transverse section; chambers embracing; wall calcareous, porcellaneous, imperforate; aperture terminal, rounded, with tooth.

Sigmoilina obesa Heron-Allen and Earland, Pl. 31, figs. 9-11.

ORIGINAL CITATION: *Sigmoilina obesa* Heron-Allen and Earland, 1932. Discovery Repts., v. 4, p. 320, pl. 7, figs. 1-4.

TYPE LOCALITY: Area of the Islas Malvinas.

AGE: Recent.

DESCRIPTION: Test compressed, triangular in transverse section; chambers convex with rounded periphery, the final occupying 3/4 of test surface; wall thick, smooth, glassy; sutures distinct; aperture small, ovate with simple, short, flat tooth.

DIAMETER: range: 400-800.

DISTRIBUTION: see Figures 11 and 17.

SIGMOMORPHINA Cushman and Ozawa, 1928

Test free, elongate, ovate, laterally compressed, chambers in megalospheric forms added in sigmoidal fashion, in microspheric forms initially quinqueloculine and later sigmoidal; chambers separated by angles somewhat less than 180°, partially embracing; wall calcareous, perforate; aperture radiate, irregularly or radially cribrate.

Sigmomorphina pauperata (Terquem), Pl. 31, figs. 12, 13.

ORIGINAL CITATION: *Polymorphina pauperata* Terquem, 1878. Mém. Soc. Géol., sér. 3, pt. 1, p. 38, pl. 3, figs. 11-19.

TYPE LOCALITY: Isle of Rhodes.

AGE: Late Pliocene.

DESCRIPTION: Test fusiform, peripheral margin rounded, ovate in transverse section, 3-4 chambers reaching base of test, the final chamber reaching only halfway to base and embracing earlier chambers; wall translucent, smooth, finely perforate and sometimes pustulose; sutures vertical in early portion, later becoming somewhat curved; aperture rounded, radiate.

LENGTH: range: 280-370.

DISTRIBUTION: see Figure 11.

Sigmomorphina williamsoni (Terquem), Pl. 31, figs. 14, 15.

ORIGINAL CITATION: *Polymorphina lactea* (Walker and

49

Jacob), var. oblonga Williamson, 1858. Foram. Gr. Brit., Ray Soc., p. 71, pl. 6, fig. 149 = *Polymorphina williamsoni* Terquem, 1878. Mém. Soc. Géol., sér. 3, pt. 1, p. 37.

TYPE LOCALITY: British Isles.

AGE: Recent.

DESCRIPTION: Test rectangular in outline, corners rounded; chambers numerous, added in alternating opposed series, leaving earlier chambers visible; wall thin, smooth, translucent; sutures curved, oblique or parallel to peripheral margin of test; aperture large, radiate, with entosolenian tube.

LENGTH: range: 280-710.

DISTRIBUTION: see Figure 11.

SPIRILLINA Ehrenberg, 1843

Test free, initially conical, later planispiral, one or both sides weakly concave; proloculus leads into undivided tubular second chamber which forms spiral of 4-9 whorls; wall calcareous, hyaline, composed of numerous calcite crystals, pores (or pseudopores) variable in size, number, and distribution from one side to other and from one specimen to another; ornamented with small shallow grooves perpendicular to spire; aperture terminal, peripheral, sometimes crecentic and turning in toward umbilicus.

Spirillina vivipara Ehrenberg, Pl. 31, figs. 16-18.

ORIGINAL CITATION: *Spirillina vivipara* Ehrenberg, 1843. Abh. Dt. Akad. Wiss. Berlin, p. 422, pl. 3, fig. 41.

TYPE LOCALITY: Vera Cruz, Mexico.

AGE: Recent.

DESCRIPTION: Test circular in outline, subconical, concavoconvex, whorls of second chamber coiled in low trochospiral providing a step-like effect toward proloculus; chambers planar on convex side, inflated on concave side; wall thin, with prominent growth lines, large pores on convex side; sutures depressed; aperture semicircular.

DIAMETER: range: 130-260.

DISTRIBUTION: see Figure 11.

SPIROLOCULINA d'Orbigny, 1826

Test free, laterally compressed, fusiform, early chambers completely enrolled about proloculus, later chambers biloculine; chambers distinct, not embracing; wall calcareous, imperforate, porcellaneous; aperture terminal with simple or bifid tooth.

Spiroloculina depressa d'Orbigny, Pl. 31, figs. 19, 20.

ORIGINAL CITATION: *Spiroloculina depressa* d'Orbigny, 1826. Ann. Sci. Nat., sér. 1, v. 7, p. 298, no. 1; Parker, Jones and Brady, 1871. Ann. Sci. Nat. Hist., ser. 4, v. 8, pl. 8, fig. 23, fig. KK.

TYPE LOCALITY: Castel Arquato, Italy.

AGE: Recent.

DESCRIPTION: Test elliptical in outline, slightly longer than wide, peripheral margin angulate; chambers markedly quadrate in section, ending in straight line where they butt against earlier chambers thus forming a zig-zag line along longitudinal axis, last chamber with neck on outside margin; wall white, shiny, rough; sutures depressed; aperture elongate, large, almost rectangular with lip and short narrow simple tooth.

LENGTH: range: 380-950.

DISTRIBUTION: see Figure 12.

Spiroloculina planulata (Lamarck), Pl. 32, figs. 1, 2.

ORIGINAL CITATION: *Miliolites planulata* Lamarck, 1805. Ann. Mus. Nat. Hist., v. 5, p. 352, no. 4 = *Spiroloculina planulata* (Lamarck), Jones, Parker and Brady, 1866. Paleontogr. Soc. Monogr., p. 15, pl. 13, figs. 37, 38.

TYPE LOCALITY: Louvres, vicinity of Paris.

AGE: Not designated.

DESCRIPTION: Test elongate; chambers quadrate with slightly rounded edges; wall white, shiny, slightly rough, very weakly striate; sutures depressed; aperture elongate, with simple long tooth.

LENGTH: range: 420-1290.

DISTRIBUTION: see Figure 12.

SPIROPLECTAMMINA Cushman, 1927

Test elongate, early portion planispiral, later biserial; chambers numerous; wall finely or coarsely agglutinate; aperture an arcuate slit at base of internal margin of last chamber.

Spiroplectammina biformis (Parker and Jones), Pl. 32, figs. 3, 4.

ORIGINAL CITATION: *Textularia agglutinans* d'Orbigny, var. biformis Parker and Jones, 1865. Roy. Soc. Lond., Philos. Trans., v. 155, p. 370, pl. 15, figs. 23, 24.

TYPE LOCALITY: West coast of Greenland.

AGE: Recent.

DESCRIPTION: Test expansion very slight, peripheral margin rounded; chambers lower than wide, inflated, 4-9 pairs; sutures depressed; wall pale yellow; aperture with narrow rim.

LENGTH: range: 160-600.

DISTRIBUTION: see Figures 12 and 17.

TEXTULARIA Defrance in de Blainville, 1824

Test free, elongate, biserial, early chambers planispiral in microspheric form, generally compressed, rarely ovate or circular in transverse section; chambers numerous; wall agglutinate, cement variable in quantity; aperture simple, interiomarginal, sometimes with lip.

Textularia agglutinans d'Orbigny, Pl. 32, figs. 5-7.

ORIGINAL CITATION: *Textularia agglutinans* d'Orbigny, 1839. In: de la Sagra, Hist. Phys. Polit. Natur. Cuba, p. 144, pl. 1, figs. 17, 18, 32-34.

TYPE LOCALITY: Cuba.

AGE: Recent.

DESCRIPTION: Test elongate, peripheral margin rounded, lobate; chambers low, wide, rounded, 8-9 pairs; wall composed of poorly sorted grains of medium size; sutures depressed, making right angle with growth axis; aperture narrow with small rim.

LENGTH: range: 380-1140.

DISTRIBUTION: see Figure 12.

Textularia candeiana d'Orbigny, Pl. 32, figs. 8-11.

ORIGINAL CITATION: *Textularia candeiana* d'Orbigny, 1839. In: de la Sagra, Hist. Phys. Polit. Natur. Cuba, p. 143, pl. 1, figs. 25-27.

TYPE LOCALITY: Cuba, Martinique and St. Thomas.

AGE: Recent.

DESCRIPTION: Test conical, early portion compressed, later portion step-like, peripheral margin acute, serrate, final pair of chambers rounded; chambers narrow in early portion, globose and convex in later portion, 11-12 pairs; wall rough, composed of diverse materials; sutures obscure, making 90° angle with growth axis; aperture very narrow, long, with small rim.

LENGTH: range: 890-1040.

DISTRIBUTION: see Figure 12.

Textularia earlandi Parker, Pl. 32, figs. 12-16.

ORIGINAL CITATION: *Textularia tenuissima* Earland, 1933. Discovery Repts., v. 7, p. 95, pl. 3, figs. 21-30 = *Textularia earlandi* Parker, 1952. Harvard Coll., Mus. Comp. Zool., Bull., v. 106, p. 458.

TYPE LOCALITY: To the south of South Georgia Island.

AGE: Recent.

DESCRIPTION: Test elongate, four times longer than wide, sometimes curved, with small apical angle, peripheral margin rounded, lobate; chambers inflated, rounded, 5-8 pairs; in some examples (microspheric?) initial coiled portion with 4-6 chambers is well developed; wall thin, pale yellow-white, composed of poorly sorted grains of medium to large size for the genus; sutures depressed making 90° angle with growth axis; aperture lunate on convex apertural face.

LENGTH: range: 210-740.

DISTRIBUTION: see Figure 12. This species also occurs in the brackish waters of Lagoa dos Patos.

Textularia gramen d'Orbigny, Pl. 32, figs. 17-21.

ORIGINAL CITATION: *Textularia gramen* d'Orbigny, 1846. Foram. Foss. Bass. Tert. Vienne, p. 248, pl. 15, figs. 4, 6.

TYPE LOCALITY: South of Vienna.

AGE: Tertiary.

DESCRIPTION: Test variable in form, subrhomboidal in transverse section, peripheral margin acute to subacute, lobate; chambers low, curved, elongate, 6-9 pairs; wall smooth, porous, composed of relatively small gray white grains mixed with some heavy minerals; sutures slightly depressed; aperture a broad slit, sometimes with rim.

LENGTH: range: 210-1070; mean: 560; coefficient of variation: 0.32.

OBSERVATIONS: This is a highly variable species, especially in the gross from of the test, the size of the final pair of chambers, the presence or absence of an apertural rim, and in the nature of the peripheral margin.

DISTRIBUTION: see Figure 12.

TRILOCULINA d'Orbigny, 1826

Test free, initial chambers in microspheric forms quinqueloculine, later triloculine separated by 120°; chambers embracing; wall calcareous, imperforate, porcellaneous, rarely with agglutinated covering; aperture terminal with tooth.

Triloculina baldai Bermúdez and Seiglie, Pl. 33, figs. 1-4.

ORIGINAL CITATION: *Triloculina baldai* Bermúdez and Seiglie, 1963. Inst. Ocean. Univ. Orient., Bol., v. 2, no. 2, p. 177, pl. 10, fig. 2.

TYPE LOCALITY: Camp. Araya, Venezuela.

AGE: Recent.

DESCRIPTION: Test subcircular in outline, massive, irregularly triangulate in transverse section, peripheral margin angulate; chambers wide; wall thick, very weakly striate, white; sutures distinct, slightly depressed; aperture ovate with thick rim and simple tooth.

LENGTH: range: 380-1150; mean: 630; coefficient of variation: 0.29.

DISTRIBUTION: see Figure 12.

Triloculina cultrata (Brady), Pl. 33, figs. 5-7.

ORIGINAL CITATION: *Miliolina cultrata* Brady, 1881. Quart. Jour. Micro. Sci., v. 21, p. 45; 1884. Challenger Exp., Repts., Zool., v. 9, pl. 5, figs. 1, 2.

TYPE LOCALITY: Figured specimens from Humboldt Bay, Papua.

AGE: Recent.

DESCRIPTION: Test elongate, compressed, peripheral margin subangulate; chambers narrow; wall thin, shiny; sutures depressed; final portion of last chamber projecting over aboral portion of penultimate chamber; aperture ovate, large, without tooth.

LENGTH: range: 190-300.

OBSERVATIONS: The absence of a tooth is a distinguishing feature of this species.

DISTRIBUTION: see Figure 12.

Triloculina laevigata (d'Orbigny), Pl. 33, figs. 8-10.
ORIGINAL CITATION: *Quinqueloculina laevigata* d'Orbigny, 1839. In: Barker-Webb and Berthelot, Hist. Nat. Canaries, v. 2, no. 2, p. 143, pl. 3, figs. 31-33.
TYPE LOCALITY: Tenerife, Canary Islands.
AGE: Recent.
DESCRIPTION: Test elongate, aboral part rounded, apertural end truncate, broadly triangulate in transverse section, peripheral margin rounded; chambers raised above surface in early whorls; wall smooth, shiny; sutures distinct; aperture ovate with simple flat tooth.
LENGTH: range: 270-630.
OBSERVATIONS: This species is distinguished from *T. oblonga* by its oblique inner chambers.
DISTRIBUTION: see Figure 12.

Triloculina oblonga (Montagu), Pl. 33, figs. 11-13.
ORIGINAL CITATION: *Vermiculum oblongum* Montagu, 1803. Testac. Brit., p. 522, pl. 14, fig. 9.
TYPE LOCALITY: Devonshire, England.
AGE: Recent.
DESCRIPTION: Test subrectangular in outline, elongate, broadly triangulate in transverse section; chambers robust with weak transverse striae; wall shiny, thick, white; sutures distinct; aperture large, rounded, with simple or bifid tooth.
LENGTH: range: 190-460.
DISTRIBUTION: see Figure 12.

Triloculina trigonula (Lamarck), Pl. 33, figs. 14-16.
ORIGINAL CITATION: *Miliolites (trigonula)* Lamarck, 1804. Ann. Mus. Nat. Hist., v. 5, p. 35; 1807. v. 9, pl. 17, fig. 4.
TYPE LOCALITY: Grignon, France.
AGE: Eocene.
DESCRIPTION: Test subcircular in outline, triangular in transverse section, peripheral margin rounded; chambers robust, convex; wall thick, white, with weak striae; sutures depressed; aperture rounded with simple or bifid tooth.
LENGTH: range: 210-380.
DISTRIBUTION: see Figure 12.

TROCHAMMINA Parker and Jones, 1879

Test free or attached, trochospiral; chambers numerous, globular or compressed; wall agglutinate; aperture interiomarginal, an arcuate slit, sometimes with lip.

Trochammina inflata (Montagu), Pl. 33, figs. 17-19.
ORIGINAL CITATION: *Nautilus inflatus* Montagu, 1803. Testac. Brit., p. 8, pl. 18, fig. 3.
TYPE LOCALITY: Coast of Devon, England.
AGE: Recent.
DESCRIPTION: Test subcircular in outline, lobate, peripheral margin rounded, 2½-3 volutions; chambers somewhat

globose, quadrangular on spiral side, triangular on umbilical side, 5-6 in final whorl, expanding regularly as added; wall thin, opaque, smooth, finely agglutinated; sutures depressed, straight on umbilical side, somewhat curved on spiral side; aperture narrow, extraumbilical, with thin lip.
DIAMETER: range: 270-670; mean: 410; coefficient of variation: 0.24.
DISTRIBUTION: see Figure 12.

Trochammina ochracea (Williamson), Pl. 33, figs. 20-22.
ORIGINAL CITATION: *Rotalina ochracea* Williamson, 1858. Foram. Gr. Brit., Ray Soc., p. 55, pl. 4, fig. 112, pl. 5, fig. 113.
TYPE LOCALITY: Shetland, British Isles.
AGE: Recent.
DESCRIPTION: Test circular in outline, somewhat lobate, compressed, peripheral margin acute, umbilical side concave with depressed umbo, sometimes partially involute, spiral side convex with 2½-3 volutions; chambers low, broad, arcuate, compressed, 8½-11 in final whorl; wall thin, smooth, yellow-white in last whorl, darker in early whorls; sutures depressed, wide, opaque, curved on spiral side, lobate, sinuous and raised when dry on umbilical side; aperture ovate, peripheral.
DIAMETER: range: 170-300; mean: 220; coefficient of variation: 0.18.
DISTRIBUTION: see Figure 12. This species also occurs in the brackish waters of Lagoa dos Patos, Rio de la Plata and Mar Chiquita.

Trochammina plana discorbis Earland, Pl. 34, figs. 1-4.
ORIGINAL CITATION: *Trochammina discorbis* Earland, 1934. Discovery Repts., v. 10, p. 104, pl. 3, figs. 28-31.
TYPE LOCALITY: Scotia Sea and Bellingshausen Sea.
AGE: Recent.
DESCRIPTION: Test circular in outline, lobate, peripheral margin rounded, spiral side moderately to strongly convex and globular with 3½ volutions, umbilical side concave; chambers inflated, triangular on umbilical side often exhibiting triangular flaps extending into umbilicus, quadrangular to arcuate on spiral side, 5-5½ in final whorl; septal sutures slightly arcuate and depressed on both sides; spiral sutures exhibiting a festooned pattern; wall somewhat rough, yellow-white, composed of moderately large grains; aperture small, located beneath cameral flap.
DIAMETER: range: 140-340.
DISTRIBUTION: see Figures 12 and 17.

Trochammina ex gr. *T. squamata* Jones and Parker, Pl. 34, figs. 5-8.
ORIGINAL CITATION: *Trochammina squamata* Jones and Parker, 1860. Quart. Jour. Geol. Soc., v. 16, p. 304; 1913. Heron-Allen and Earland, Roy. Irish Acad., Proc. v. 31,

no. 64, p. 50, pl. 3, figs. 7-10.
TYPE LOCALITY: Isle of Crete.
AGE: Recent.
DESCRIPTION: Test circular in outline, relatively strongly compressed, somewhat lobate, spiral side convex, umbilical side concave with deep umbilicus, peripheral margin rounded, 2-3 volutions; chambers quadrangular to curved on spiral side, kidney shaped on umbilical side and terminating with triangular lobe in umbilicus, 6-10 in final whorl; wall thin, smooth, finely agglutinate, ochre in final whorl, darker in earlier whorls; sutures depressed; aperture a narrow umbilical – extraumbilical slit.
DIAMETER: range: 170-360.
OBSERVATIONS: This is a highly variable species, particularly with regard to the number and shape of the chambers and the nature of the umbilicus and sutures. Detailed studies of large populations of *T. squamata* are necessary to clarify the taxonomic status of this group of forms. At the moment we prefer to designate this population as *T. ex gr. T. squamata* following the morphological interpretations of Heron-Allen and Earland (loc. cit.).
DISTRIBUTION: see Figure 12.

TUBINELLA Rhumbler, 1906

Test elongate, proloculus bulbous, second chamber appressed and growing in opposite direction, later chambers uniserial; chambers cylindrical; septa partial, delicate, visible in transmitted light; wall calcareous, imperforate, porcellaneous; aperture terminal, simple.

Tubinella funalis (Brady), Pl. 34, figs. 9-11.
ORIGINAL CITATION: *Articulina funalis* Brady, 1884. Challenger Exp.. Repts.. Zool.. v. 9, p. 185. pl. 13. figs. 6-11.
TYPE LOCALITY: Kerguelen Island. Prince Edward Island. Papua Island.
AGE: Recent.
DESCRIPTION: Test straight or irregularly arcuate, initial chamber an elongate inflated bulb, additional chambers variable in height and diameter; wall covered with fine, parallel, longitudinal costae, increasing in number toward apertural end by bifurcation; sutures obscure, perpendicular to growth axis; aperture simple, sometimes slightly constricted.
LENGTH: range: 550-1430.
DISTRIBUTION: see Figure 12.

UVIGERINA d'Orbigny, 1826

Test free, elongate, generally triserial, sometimes biserial in final portion, rounded in transverse section; chambers inflated; wall calcareous, radial, perforate, smooth, hispid or costate; aperture terminal, surrounded by imperforate neck, with phialine lip, and alate tooth plate connecting with earlier chambers.

Uvigerina bifurcata d'Orbigny, Pl. 34, figs. 12-14.
ORIGINAL CITATION: *Uvigerina bifurcata* d'Orbigny, 1839. Voy. Amér. Mérid., v. 5, pt. 5, p. 53, pl. 5, fig. 113.
TYPE LOCALITY: Islas Malvinas.
AGE: Recent.
DESCRIPTION: Test of variable length, uniform width throughout most of length, lobate; chambers strongly inflated. wider than tall: 5-8 volutions: wall opaque. densely covered with thick. protruding often bifurcating costae. interrupted at sutures; sutures depressed; aperture rounded, small, on short neck, with well developed tooth.
LENGTH: range:380-1560; mean: 970; coefficient of variation: 0.24.
DISTRIBUTION: see Figure 17.

Uvigerina peregrina Cushman, forma parvula, Pl. 34, figs. 15, 16.
ORIGINAL CITATION: *Uvigerina peregrina* Cushman, var. parvula Cushman. 1923. U.S. Nat. Mus., Bull., 104, p. 168, pl. 42, fig. 11.
TYPE LOCALITY: Gulf of Mexico.
AGE: Recent.
DESCRIPTION: Test ovate in outline, tapering at ends, massive, lobate, 3-4 volutions; chambers low, weakly inflated; wall translucent to opaque, finely perforate, with high, thin costae which are interrupted at sutures, inner-costate areas and extremities of test covered with small granules and spines; sutures depressed; aperture small, on short neck with thick phialine lip.
LENGTH: range: 340-440.
DISTRIBUTION: see Figure 12.

Uvigerina striata d'Orbigny, Pl. 34, figs. 17, 18.
ORIGINAL CITATION: *Uvigerina striata* d'Orbigny, 1839. Voy. Amér. Mérid., v. 5, pt. 5, p. 53, pl. 7, fig. 16.
TYPE LOCALITY: Islas Malvinas.
AGE: Recent.
DESCRIPTION: Test acuminate at both ends, lobate, tapering toward base; 3-4 volutions; chambers high; wall translucent to opaque, thin, densely perforate, irregularly costate; costae broken at sutures; sutures depressed; aperture small with thick lip, short neck and well developed tooth.
LENGTH: range: 360-780.
OBSERVATIONS: Well preserved examples of this species are found in the original collections of d'Orbigny in Paris.
DISTRIBUTION: see Figures 12 and 17.

VIRGULINA d'Orbigny, 1826

Test elongate, fusiform, rounded or ovate in transverse section, initial portion triserial and torted, later

53

becoming biserial; wall calcareous, finely perforate, granular, smooth; sutures depressed, oblique; aperture elongate, comma shaped, extending from base of apertural face to terminal position, with tooth plate.

Virgulina riggii Boltovskoy, Pl. 34, figs. 19-22.
ORIGINAL CITATION: *Virgulina riggii* Boltovskoy, 1954. Inst. Nac. Invest. Cienc. Nat., Rev., Geol., v. 3, no. 3, p. 186, pl. 11, figs. 7-11, 15.
TYPE LOCALITY: Golfo San Jorge, Argentina.
AGE: Recent.
DESCRIPTION: Test elongate, lobate; 3-4 volutions; chambers high, weakly convex, irregularly arranged; wall smooth, thin, translucent, finely and densely perforate, sometimes iridescent; sutures perpendicular to growth axis, strongly depressed; aperture large with broad base.
LENGTH: range: 160-400; mean: 240; coefficient of variation: 0.29.
DISTRIBUTION: see Figures 12 and 17.

HYPOHALINE AND FRESH WATER SPECIES

Ammonia ? *veneta* (Schultze), Pl. 35, figs. 1-3.
ORIGINAL CITATION: *Rotalia veneta* Schultze, 1854. Organ. Polyth., p. 59, pl. 3, figs. 1-5, pl. 7, figs. 22-24.
TYPE LOCALITY: Not designated.
AGE: Recent.
DESCRIPTION: Test small, subcircular in outline, compressed, peripheral margin rounded, somewhat lobate, spiral side somewhat convex; umbilicus covered with small pustules; chambers slightly inflated, 7-8 in final whorl; wall thin, transparent, finely and densely perforate; pores in clusters of 3-5; sutures limbate, arcuate and depressed on spiral side, depressed and radial on umbilical side; aperture lunate.
DIAMETER: range: 100-230.
OBSERVATIONS: There are but a few specimens of this species and no sections were made to determine if it is truly an *Ammonia*.
DISTRIBUTION: Río Quequén. Living specimens are found in salinities between 1.9-30.3‰.

AMMOSCALARIA Höglund, 1947

Test compressed, elongate, early portion planispiral, later uniserial and rectilinear; chambers separated by delicate organic membranes with foramina which differ from the aperture; wall agglutinaed; sutures oblique, obscure; aperture terminal, narrow and delicate.

Ammoscalaria pseudospiralis (Williamson), Pl. 35, figs. 4-7.
ORIGINAL CITATION: *Proteonina pseudospiralis* Williamson, 1858. Foram. Gr. Brit., Ray Soc., p. 2, pl. 1, figs. 2, 3.

TYPE LOCALITY: Isle of Skye, Scotland.
AGE: Recent.
DESCRIPTION: Test compressed, weakly lobate, last few chambers quadrangular; wall rugose, composed of a variety of large grains; sutures obscure; final aperture elongate, narrow, irregular and often difficult to see; septal aperture round, located in central part of internal septa, with distinct bordering rim.
LENGTH: range: 720-2200.
DISTRIBUTION: This species occurs in the hypohaline waters of southern Brazil, Río Santa Lucía, Río de la Plata and Mar Chiquita.

Ammoscalaria tenuimargo (Brady), Pl. 35, figs. 8-11.
ORIGINAL CITATION: *Haplophragmium tenuimargo* Brady, 1882. Roy. Soc. Edinburgh, Proc., v. 11, p. 715; 1884. Challenger Exp., Repts., Zool., v. 9, pl. 33, figs. 13-16.
TYPE LOCALITY: 59°37′N; 7°19′W.
AGE: Recent.
DISTRIBUTION: Test J-shaped, very compressed; chambers broad, weakly inflated, 5-6 in uncoiled portion; wall coarsely agglutinate, rough; aperture simple, irregular.
LENGTH: range: 780-920.
DISTRIBUTION: Río de la Plata.

AMMOTIUM Loeblich and Tappan, 1953

Test oval to elongate in outline, compressed, early portion planispiral and evolute, later unrolled; chambers numerous, wide, low, reaching backward towards coil at inner margin; wall agglutinate; aperture terminal, simple, rounded or elongate.

Ammotium cassis (Parker), Pl. 35, figs. 12, 13.
ORIGINAL CITATION: *Lituola cassis* Parker, 1870. In: Dawson, Canadian Nat., n.s., v. 5, p. 177, textfig. 5.
TYPE LOCALITY: Gaspé Bay, Canada.
AGE: Recent.
DESCRIPTION: Test elongate, compressed, peripheral margin rounded; chambers obliquely arranged in unrolled portion, 6-7; wall smooth, yellow-white, regularly agglutinate; sutures slightly depressed, oblique and curved in uncoiled portion; aperture short, narrow.
LENGTH: range: 210-370.
DISTRIBUTION: Lagoa dos Patos.

Ammotium salsum (Cushman and Brönnimann), Pl. 35, figs. 14-17.
ORIGINAL CITATION: *Ammobaculites salsus* Cushman and Brönnimann, 1948. Cushman Lab. Foram. Res., Contr., v. 24, pt. 1, p. 16, pl. 3, figs. 7-9.
TYPE LOCALITY: West coast of Trinidad.
AGE: Recent.
DESCRIPTION: Test elongate, peripheral margin rounded, final chambers elongate, somewhat inflated; wall com-

54

posed of variety of poorly sorted grains; sutures obscure, depressed, slightly oblique in unrolled portion; aperture ovate, relatively large.

LENGTH: range: 280-480.

OBSERVATIONS: *A. salsum* differs from *A. cassis* in its less compressed test, fewer and more inflated chambers and abrupt change from coiled to uncoiled stage.

DISTRIBUTION: The species is typical of hypohaline waters in the northern part of the area of study (Lagoa dos Patos, Lagoa Mirim, Arroio Chuí, Río Santa Lucía, and Río de la Plata).

ARENOPARRELLA Andersen, 1951

Test free, trochospiral, spiral side convex; umbilicus small, closed; wall agglutinated, thin, generally smooth; aperture an elongate slit rising from basal suture up apertural face parallel to coiling direction; supplementary apertures cribrate in center of apertural face.

Arenoparrella mexicana (Kornfeld), Pl. 35, figs. 18-21.

ORIGINAL CITATION: *Trochammina inflata* (Montagu), var. mexicana Kornfeld, 1931. Stanford Univ., Dept. Geol., Contr., v. 1, p. 86, pl. 13, fig. 5.

TYPE LOCALITY: Timbalier Island, Louisiana, USA.

AGE: Recent.

DESCRIPTION: Test ovate to subcircular in outline, compressed, peripheral margin rounded, weakly lobate, spiral side evolute with 3 volutions, umbilical side involute; chambers quadrangular on spiral side, triangular on umbilical side, 5-5½ in final whorl; wall finely agglutinate, smooth, somewhat shiny, ochre; sutures somewhat depressed, straight on spiral side, somewhat curved on umbilical side; principal aperture straight and offset toward spiral side; supplementary aperture an ovate hole.

LENGTH: range: 280-380.

DISTRIBUTION: This species occurs in several hypohaline areas in the northern part of the study area.

HAPLOPHRAGMOIDES Cushman, 1910

Test planispiral, involute; chambers somewhat inflated, variable in number; wall agglutinated; aperture an interiomarginal, equatorial slit.

Haplophragmoides wilberti Andersen, Pl. 36, figs. 1-4.

ORIGINAL CITATION: *Haplophragmoides wilberti* Andersen, 1953. Cushman Found. Foram. Res., Contr., v. 4, p. 21, pl. 4, fig. 7.

TYPE LOCALITY: Dog Lake, Louisiana, USA.

AGE: Recent.

DESCRIPTION: Test circular in outline, weakly lobate, peripheral margin rounded; 6-8 triangular chambers in final whorl; wall smooth, shiny, finely agglutinate with much

cement, yellow-gray; sutures slightly depressed, straight to sigmoid; aperture a small arc with narrow projecting lip.

DIAMETER: range: 170-360.

DISTRIBUTION: Lagoa dos Patos and Río Quequén where it lives in waters of salinity as low as 0.5‰. It also occurs in fresh water sites which are periodically invaded by brackish water.

JADAMMINA Bartenstein and Brand, 1938

Test free, trochospiral, flattened, lobate, peripheral margin rounded to subacute, spiral side with depressed umbilicus; chambers broader than high, 8-9 in final whorl; wall imperforate, smooth, finely agglutinate over an organic membrane which tends to collapse when dry, gray-yellow-white; sutures depressed and slightly sigmoid; aperture interiomarginal with supplementary apertures areal, ovate, with lip, variable in number and location.

Jadammina polystoma Bartenstein and Brand, Pl. 36, figs. 5-7.

ORIGINAL CITATION: *Jadammina polystoma* Bartenstein and Brand, 1938. Senckenbergiana, v. 20, no. 5, p. 381, textfigs. 1-3.

TYPE LOCALITY: Jade Bay, northwest Germany.

AGE: Recent.

DESCRIPTION: *Jadammina* is monospecific and its description corresponds to that of the genus.

DIAMETER: range: 240-600.

DISTRIBUTION: Río Quequén, Puerto Deseado and Isla de los Estados.

MILIAMMINA Heron-Allen and Earland, 1930

Test triloculine or quinqueloculine; wall agglutinated, thin; aperture terminal, rounded, sometimes with simple tooth.

Miliammina fusca (Brady), Pl. 36, figs. 8-12.

ORIGINAL CITATION: *Quinqueloculina fusca* Brady, 1870. Ann. Mag. Nat. Hist., ser. 4, v. 6, p. 286, pl. 11, figs. 2, 3.

TYPE LOCALITY: Not designated.

AGE: Recent.

DESCRIPTION: Test ovate, aboral part rounded, triangular in transverse section, chambers elongate, maintaining their width throughout length; wall finely and uniformly agglutinate on an organic matrix, weak transverse undulations, dark gray to light brown; sutures depressed; aperture ovate to lunate sometimes with short flat tooth.

LENGTH: range: 230-700; mean: 410; coefficient of variation: 0.29.

OBSERVATION: This species is euryhaline and is quite typical of lagoons, estuaries and marshes where the salinity is less than normal. In the study area the majority of the specimens occur in water whose salinity lies between 8-20‰. Those specimens lying near the extremes of this range are usually smaller and more likely to be pathological. Only rarely is *M. fusca* found in fresh water sites that are periodically invaded by brackish water.

DISTRIBUTION: This cosmopolitan species occurs in practically all hypohaline areas.

Nonion ? pseudotisburyense Boltovskoy and Giussani, Pl. 36, figs. 13-19.

ORIGINAL CITATION: *Nonion? pseudotisburyense* Boltovskoy and Giussani, in press.

TYPE LOCALITY: Delta of the Río Paraná, Argentina.

AGE: Recent.

DESCRIPTION: Test free, circular in outline, lobate, particularly in adult forms, peripheral margin rounded, generally symmetrical, evolute with 1-2 whorls; chambers inflated, increasing slowly in size as added, 6-9 in final whorl of adult forms; proloculus large, circular, pustulose; sutures depressed, open near umbilicus, with pustules along margins; wall very thin, smooth, with large, uniformily distributed pores; apertural face rounded, somewhat convex, covered with pustules; aperture, formed of several round openings at base of apertural face, often obscured by pustules; supplementary apertures consist of irregularly rounded holes of varying diameter scattered on the central portion of the apertural face, each surrounded by small pustules.

DIAMETER: range: 160-330; mean: 240; coefficient of variation: 0.21.

OBSERVATIONS: This is the first calcareous, multilocular foraminifer found in great quantity living in completely fresh water. It was found initially in the Río Paraná (Boltovskoy, 1958a) and identified with some reservation as *N. tisburyense*. It was later found living in the fresh and brackish waters of Lagoa dos Patos and Lagoa Mirim. (Closs and Medeiros, 1965, 1967; Madeira-Falcetta, 1974) where it was also identified as *N. tisburyense*. These South American occurences appear to be unique as, to our knowledge, this form has not been found elsewhere living in completely fresh water.

The generic status of this species is unclear. The presence of irregular supplementary apertures, and the division of the basal aperture into several openings set this form apart from *Nonion* although *Nonion* seems to be the most closely related genus.

DISTRIBUTION: Fresh and hypohaline waters of the Río Paraná, Río de la Plata and Lagoa Mirim.

PROTOSCHISTA Eimer and Fickert, 1899

Test free, uniserial or branching from proloculus to form 2-3 uniserial arms; chambers slightly inflated, equal in size; wall agglutinated, little cement, rugose; aperture terminal, circular.

Protoschista findens (Parker), Pl. 36, figs. 20, 21.

ORIGINAL CITATION: *Lituola findens* Parker, 1870. In: Dawson, Canadian Nat., n.s., v. 5, p. 176, 177, textfig. 1.

TYPE LOCALITY: Not designated.

AGE: Recent.

DESCRIPTION: Test elongate, circular in transverse section; chambers low, wide, up to 14 per test; wall pale yellow, composed of medium sized grains; sutures lying at 90° angle to growth axis, slightly depressed; aperture central, large.

LENGTH: range: 330-890.

DISTRIBUTION: Lagoa dos Patos.

56

REFERENCES

Berggren, W. A. and Hollister, C. D., 1974. Paleogeography paleobiogeography and the history of circulation in the Atlantic Ocean. In: Hay, W. W., 1974, Studies in Paleo-oceanography. Soc. Econ. Paleontol. Mineral., Sp. Publ. 20: 126-186.

Boltovskoy, E., 1953. Über Zersetzungserscheinungen bei mikro-paläontologische Sammlungsmaterial: Paläont. Z., v. 27, n° 3/4, p. 237.240.

Boltovskoy, E., 1954a. Beobachtungen über Einfluss der Ernährung auf die Foraminiferenschalen: Paläont. Z., v. 28, n° 3/4, p. 204-207.

Boltovskoy, E., 1954b. Foraminíferos del golfo San Jorge: Inst. Nac. Invest. Cienc. Nat., Rev., Geol., v. 3, n° 3, p. 85-246.

Boltovskoy, E., 1954c. Foraminíferos de la bahía San Blas: Inst. Nac. Invest. Cienc. Nat., Rev., Geol., v. 3, n° 4, p. 247-300.

Boltovskoy, E., 1955. Recent foraminifera from shore sand at Quequén, province of Buenos Aires, and change in the foraminiferal fauna to the north and south: Cushman Found. Foram. Res., Contr., v. 6, n° 1, p. 39-42.

Boltovskoy, E., 1956a. Applications of chemical ecology in the study of foraminifera: Micropaleontology, v. 2, n° 4, p. 321-325.

Boltovskoy, E., 1956b. On the cyclical occurrence of foraminifera: Dusenia, v. 7, n° 4, p. 211-218.

Boltovskoy, E., 1957a. Las anormalidades en los caparazones de foraminíferos y el "Indice de regeneramiento": Ameghiniana, v. 1, n° 1/2, p. 80-84.

Boltovskoy, E., 1957b. Los foraminíferos del estuario del Río de la Plata y su zona de influencia: Inst. Nac. Cienc. Nat., Rev., Geol., v. 6, n° 1, p. 1-76.

Boltovskoy, E., 1958a. The foraminifera fauna of the Río de la Plata and its relation to the Caribbean area: Cushman Found. Foram. Res., Contr., v. 9, n° 1, p. 17-21.

Boltovskoy, E., 1958b. Problems in taxanomy and nomenclature exemplified by *Nonion affine* (Reuss): Micropaleontology, v. 4, n° 2, p. 193-200.

Boltovskoy, E., 1959a. Foraminifera as biological indicators in the study of ocean currents: Micropaleontology, v. 5, n° 4, p. 473-481.

Boltovskoy, E., 1959b. La corriente de Malvinas (un estudio en base a la investigación de foraminíferos): Argentina, Serv. Hidr. Nav., H. 1015, p. 1-96.

Boltovskoy, E., 1961. Problemas de ecología química en la Argentina: Cienc. Invest., v. 17, n° 4, p. 97-111.

Boltovskoy, E., 1962. Foraminíferos de la plataforma continental entre el cabo Santo Tomé y la desembocadura del Río de la Plata: Mus. Argent. Cienc. Nat., Rev., Zool., v. 6, n° 6, p. 249-346.

Boltovskoy, E., 1963a. The littoral foraminiferal biocoenoses of Puerto Deseado (Patagonia, Argentina): Cushman Found. Foram. Res., Contr. v. 14. n° 2, p. 58-70.

Boltovskoy, E., 1963b. Sobre las relaciones entre foraminíferos y turbelarios: Neotrópica, v. 9, n° 29, p. 55-60.

Boltovskoy, E., 1963c. Foraminiferos y sus relaciones con el medio: Mus. Argent. Cienc. Nat., Rev., Hidrobiol., v. 1, n° 2, p. 21-109.

Boltovskoy, E., 1964. Provincias zoogeográficas de América del Sur y su sector Antártico, según los foraminíferos bentónicos: Inst. Biol. Mar. Mar. del Plata, Bol., 7, p. 93-98.

Boltovskoy, E., 1965a. Los Foraminíferos Recientes (biología, métodos de estudio, aplicación oceanográfica): EUDEBA, Buenos Aires, 510 pp.

Boltovskoy, E., 1965b. Recolección de foraminíferos en las aguas someras y su preparación: Centr. Inv. Biol. Mar., Contr. Tecn. n° 1, p. 1-11.

Boltovskoy, E., 1965c. Beitrag zur Kenntnis der Jahreszyklen der Foraminiferen: Int. Rev. ges. Hydrobiol., v. 50, n° 2, p. 203-296.

Boltovskoy, E., 1965d. Twilight of foraminiferology: Jour. Paleont., v. 39, n° 3, p. 383-390.

Boltovskoy, E., 1966. Depth at which foraminifera can survive in sediments: Cushman Found. Foram. Res., Contr., v. 17, n° 17, n° 2, p. 43-45.

Boltovskoy, E., 1967. Indicadores biológicos en la oceanografia: Cienc. Invest., v. 23, n° 2, p. 66-75.

Boltovskoy, E., 1969. Foraminifera as hydrological indicators: In: Brönnimann, P. & Renz, H. H. (eds.), Proc. Ist. Int. Conf. Plankt. Microfoss., v. 2, p. 1-14, Brill, Leiden.

Boltovskoy, E., 1970a. Distribution of the marine littoral foraminifera in Argentina, Uruguay and Southern Brazil: Mar. Biol., v. 6, n° 4, p. 335-344.

Boltovskoy, E., 1970b. Masas de agua (característica, distribución, movimientos) en la superficie del Atlántico Sudoeste, según indicadores biológicos – foraminíferos: Argentina, Serv. Hidr. Nav., H. 643, p. 1-99.

Boltovskoy, E., 1971. Relationship between benthonic foraminiferal fauna and the substrate in the littoral zone: Jour. Mar. Geol. (Japan), v. 7, n° 1, p. 26-30.

Boltovskoy, E., 1973. Estudio de testigos submarinos del Atlántico Sudoccidental: Mus. Argent. Cienc. Nat., Rev., Geol., v. 7, n° 4, p. 215-340.

Boltovskoy, E., 1976. Distribution of Recent foraminifera of the South American Region: In: Hedley, R. H. & Adams, C. G. (eds.), Foraminifera, v. 2, p. 171-236, Academic Press, London.

Boltovskoy, E. & Boltovskoy, A., 1968. Foraminíferos y Tecamebas de la parte interior del Río Quequén Grande (sistemática, distribución, ecología): Mus. Argent. Cienc. Nat., Rev., Hidrobiol., v. 2, n° 4, p. 127-164.

Boltovskoy, E. & Giussani, G., in press. *Nonion? pseudotisburyense* n. sp. – Primer foraminífero calcáreo multilocular hallado en aguas de río.

Boltovskoy, E. & Lena, H., 1966. Unrecorded foraminifera from the littoral of Puerto Deseado: Cushman Found. Foram. Res., Contr., v. 17, n° 4, p. 144-149.

Boltovskoy, E. & Lena, H., 1969a. Seasonal occurrences, standing crop and production in benthic foraminifera of Puerto Deseado: Cushman Found. Foram. Res., Contr., v. 20, p. 87-93.

Boltovskoy, E. & Lena, H., 1969b. Los epibiontes de *Macrocystis* flotante como indicadores hidrológicos: Neotrópica, v. 15, n° 48, p. 135-137.

Boltovskoy, E. & Lena, H., 1969c. Microdistribution des foraminifères benthoniques vivants: Rev. Micropal., v. 12, n° 3, p. 177-185.

Boltovskoy, E. & Lena, H., 1970. Additional note on unrecorded foraminifera from littoral of Puerto Deseado (Patagonia, Argentina): Cushman Found. Foram. Res., Contr., v. 21, n° 4, p. 148-155.

Boltovskoy, E. & Lena, H., 1971. The Foraminifera (except family Allogromiidae) which dwell in fresh water: Jour. Foram. Res., v. 1, n° 2, p. 71-76.

Boltovskoy, E. & Lena, H., 1974. Foraminíferos del Río de la Plata: Argentina, Serv. Hidr. Nav., H. 661, p. 1-22.

Boltovskoy, E. & Wright, R., 1976. Recent Foraminifera: Dr. W. Junk b.v., The Hague, 515 pp.

Brady, H. B., 1884. Report of the foraminifera dredged by H.M.S. "Challenger" during the years 1873-1876: Rept. Voy. Challenger, Zool., v. 9, p. 1-814.

Closs, D., 1962. Foraminíferos e Tecamebas de Lagoa dos Patos (R. G. S.): Esc. Geol. Porto Alegre, Bol. n° 11, p. 1-130.

Closs, D. & Barberena, M. C., 1962a. Foraminíferos recentes das praias do litoral sul-Brasileiro. 1. Arroio Chuí (R. G. S.). Araranguá (S. C.): Univ. Rio Grande do Sul, Bol. I.C.N., n° 16, p. 7-55.

Closs, D. & Barberena, M. C., 1962b. Faunal studies of Recent foraminifera from the shore sands of the state Rio Grande do Sul in southern Brazil: Cushman Found. Foram. Res., Contr., v. 13, p. 74-78.

Closs, D. & Madeira, M., 1962. Tecamebas e foraminíferos do Arroio Chuí (Santa Vitoria do Palmar, R. Grande do Sul, Brazil): Iheringia, Zool., n° 19, p. 1-44.

Closs, D. & Madeira, M., 1968. Seasonal variations of brackish foraminifera in the Patos Lagoon, Southern Brazil: Univ. R. G. S., Esc. Geol., Publ. Esp. 15, p. 1-51.

Closs, D. & Medeiros, J. M., 1965. New observations on the ecological subdivision of the Patos Lagoon in Southern Brazil: Univ. R. G. S., Inst. Cienc. Nat., Bol. 24, p. 3-33.

Closs, D. & Medeiros, V. M., 1967. Thecamoebina and foraminifera from the Mirim Lagoon, southern Brazil: Iheringia, Zool., n° 35, p. 75-88.

Cushman, J. A. & Parker, F. L. 1931. Recent foraminifera from the Atlantic coast of South America: U.S. Nat. Mus., Proc., v. 80, art. 3, p. 1-24.

Egger, J. C., 1893. Foraminiferen aus Meeresgrundproben gelothet von 1874-1876 von S. M. Sch. "Gazelle": K. Bayer. Akad. Wiss., Abhandl., Math-Phys. Cl., v. 18, p. 193-458.

Forti, I. R. S. & Roettger, E. U., 1967. Further observation on the seasonal variations of mixohaline foraminifera from the Patos Lagoon, Southern Brazil: Arch. Ocean. Limmol., 15, p. 55-61.

Frenguelli, J., 1935. "*Silicotextulina deflandrei*"nueva especie de foraminífero siliceo viviente en el Puerto de San Blas (Provincia de Buenos Aires): Mus. La Plata, v. 1, Zool., 1, p. 113-119.

Frenguelli, J., 1947. *Silicotextulina melchersi* Freng. en el placton del Puerto Quequén: Mus. La Plata, Not., v. 11, Zool., n° 96, p. 345-347.

Giussani, G. & Watanabe, S., in press. Foraminíferos bentónicos como indicadores de la corriente de Malvinas.

Hansen, H. J., & Lykke-Andersen, A., 1976. Wall structure and classification of fossil and recent elphidiid and nonionid Foraminifera: Fossils and Strata, No. 10, 37 pp.

Herb, R. 1971. Distribution of Recent benthonic foraminifera in the Drake Passage: In: Biology of the Antarctic Seas IV; Antarct. Res. Ser., 17, p. 251-300.

Heron-Allen, E. & Earland, A., 1932. Foraminifera. Pt. 1. The ice free area of the Falkland Islands and adjacent seas: Discovery Rep., v. 4, p. 291-460.

Kihle, R. & Lofaldli, M. 1973. Atlas of Foraminifera from Unconsolidated Sediments on the Norwegian Continental Shelf. Description and Stratigraphic Occurrence of 214 Species: Nor. Tek.-Naturvitensk. Forskningsrad, Cont. Shelf Div., Cont. Shelf Proj., Publ. 35, unpaginated.

Lena, H., 1966. Foraminíferos recientes de Ushuaia (Tierra del Fuego, Argentina). Ameghiniana, v. 4, n° 9, p. 311-322.

Lena, H., 1972. Cytological studies in *Allogromia flexibilis* (Wiesner). Int. Rev. ges. Hydrobiol., v. 57, n° 4, p. 637-644.

Lena, H., 1974. *Dahlgrenia patagoniensis* gen. nov., sp. nov. (Foraminifera, Saccamminidae). Physis, Secc. A, v. 33, n° 86, p. 9-16.

Lena, H., 1976. Distribución de los foraminíferos bentónicos en el área oceánica adyacente al Río de la Plata. Physis, Secc. A, v. 35, n° 91, p. 135-144.

Lena, H. & Freire, F., 1974. Estudios citológicos en *Allogromia laticollaris* Arnold. Physis, Secc. A, v. 33, n° 86, p. 123-133.

Lena, H. & L'hoste, S. G., 1975. Foraminíferos de aguas salobres (Mar Chiquita, Argentina). Rev. Españ. Micropaleont., v. 7, n° 3, p. 539-548.

Le Calvez, Y., 1969. Remarques sur la conception et la taxinomie de quelques genres de Foraminifères. Cah. Micropaléont., v. 1, n° 13, p. 1-13.

Le Calvez, Y., 1974. Révision de foraminifères de la collection d'Orbigny. I-Foraminifères des Îles Canaries. Cah. Micropaléont., C.N.R.S., p. 1-107.

Le Calvez, Y., 1977. Revision des foraminifères de la collection d'Orbigny. II-Foraminifères de l'Île de Cuba. Cah. MIcropaléont. C.N.R.S., v. 1, p. 1-128, v. 2, p. 1-130.

Loeblich, A. R. & Tappan, H., 1964. In: Treatise on invertebrate paleontology (ed. R. C. Moore), Pt. C, Protista 2, Sarcodina (chiefly "Thecamoebians" and Foraminifera), Geol. Soc. Amer. & Univ. Kansas, 900 pp.

Loeblich, A. R. & Tappan, H., 1974. Recent advances in the classification of the Foraminiferida. In: R. H. Hedley & C. G. Adams (eds.). Foraminifera, v. 1, p. 1-53, Academic Press, London.

Madeira-Falcetta, M., 1974. Ecological distribution of the thecamoebal and foraminiferal associations in the mixohaline environments of the southern Brazilian littoral. An. Acad. Brasil. Cienc., v. 46, n° 3/4, p. 667-687.

Murray, J., 1895. Summary of the scientific results obtained at the sounding, dredging, and trawling stations of H. M. S. "Challenger": Challenger Repts., Summary, v. 2, p. 797-1608.

Murray, J. W., 1971. An Atlas of British Recent Foraminiferids. Heinemann, London, 244 pp.

Orbigny, A. D., 1839. Voyage dans l'Amérique Méridionale: Foraminifères, v. 5, p. 1-86, (atlas, v. 9, 1847).

Pearcy, F. G., 1914. Foraminifera of the Scottish National Antarctic Expedition: Trans. Roy. Soc. Edinbourgh, v. 49, n° 4, p. 991-1044.

Scarabino, V., 1967. Ecología de foraminíferos del Río Santa Lucia (Dpto. de Montevideo, Uruguay): Inst. Invest. Pesq., Rev., v. 2, n° 1, p. 139-161.

Theyer, F., 1966. Variationstatistische Untersuchungen zur Verbreitung der Gattung *Buccella* Anderson im Südlichen Teil Südamerikas (Protozoa, Foraminifera): Zool. Jahrb. Abt. Syst. Ökol. Geogr. Tiere, v. 93, no. 2, p. 203-222.

Thompson, L. B., 1978. Distribution of living benthonic foraminifera, Isla de los Estados, Tierra del Fuego, Argentina: Jour. Foram. Res., v. 8, n° 3, p. 241-257.

Todd, R., 1963. Nomenclature of Foraminifera: Cushman Found. For. Res., Contr., v. 14, n° 3, p. 109-111.

Williamson, W. C., 1858. On the Recent Foraminifera of Great Britain. Ray Soc., London, 107 pp.

Wright, R. C., 1968. Miliolidae (foraminíferos) recientes del estuario del río Quequén Grande (Prov. de Bs. As.): Mus. Argent. Cienc. Nat., Rev., Hidrobiol., v. 2, n° 7, p. 225-256.

PLATES

Plate 1

Fig. 1, *Allogromia flexibilis*, × 60; Puerto Deseado

Fig. 2, *Allogromia flexibilis*, × 60; Puerto Deseado

Fig. 3, *Allogromia flexibilis*, × 500; aperture; Puerto Deseado

Fig. 4, *Ammonia beccarii*, × 80, spiral side; Mar del Plata

Fig. 5, *Ammonia beccarii*, × 200, umbilical side; Mar del Plata

Fig. 6, *Ammonia beccarii*, × 70, edge view; south of Golfo San Jorge

Fig. 7, *Ammonia beccarii*, × 250, detail of wall showing tubercules; south of Golfo San Jorge

Fig. 8, *Ammonia* ex gr. *A. parkinsoniana*, × 100, spiral side; Río Quequén

Fig. 9, *Ammonia* ex gr. *A. parkinsoniana*, × 125, umbilical side; Río Quequén

Fig. 10, *Amphicoryna scalaris*, × 150; Bahía San Blas

Fig. 11, *Amphicoryna scalaris*, × 100; Golfo San Jorge

Fig. 12, *Amphicoryna scalaris*, × 200; Golfo San Jorge

Fig. 13, *Angulogerina angulosa angulosa*, × 70; Río de la Plata

Fig. 14, *Angulogerina angulosa angulosa*, × 80; 39°58′S, 56°57′W

Fig. 15, *Angulogerina angulosa angulosa*, × 200, apertural view; Río de la Plata

Fig. 16, *Angulogerina angulosa angulosa*, × 700, detail of wall showing pores and costae; Río de la Plata.

Fig. 17, *Angulogerina angulosa occidentalis*, × 180; Río de la Plata

Fig. 18, *Angulogerina angulosa occidentalis*, × 145; Río de la Plata

Fig. 19, *Anomalina vermiculata*, × 75, dorsal side; Malvin Current, 49°S

Fig. 20, *Anomalina vermiculata*, × 30, ventral side; Islas Malvinas, (probably topotype)

Fig. 21, *Anomalina vermiculata*, × 35, edge view; Islas Malvinas, (probably topotype)

61

Plate 2

Fig. 1, *Astacolus crepidulus*, × 75; southern coast of Brazil

Fig. 2, *Astacolus crepidulus*, × 160, edge view; southern coast of Brazil

Fig. 3, *Asterigerinata pacifica*, × 200, spiral side; Puerto Deseado

Fig. 4, *Asterigerinata pacifica*, × 100, umbilical side; Puerto Deseado

Fig. 5, *Asterigerinata pacifica*, × 300, edge view; Puerto Deseado

Fig. 6, *Biloculinella irregularis*, × 100, frontal view; Islas Malvinas (topotype)

Fig. 7, *Biloculinella irregularis*, × 100, side view; Islas Malvinas (topotype)

Fig. 8, *Biloculinella irregularis*, × 100, apertural view; Islas Malvinas (topotype)

Fig. 9, *Bolivina compacta*, × 150; Bahía San Blas

Fig. 10, *Bolivina compacta*, × 110; Bahía San Blas

Fig. 11, *Bolivina* cf. *B. danvillensis*, × 135; Bahía San Blas

Fig. 12, *Bolivina compacta*, × 200, oblique view of aperture and peripheral margin; Bahía San Blas

Fig. 13, *Bolivina* cf. *B. danvillensis*, × 375, peripheral margin; Bahía San Blas

Fig. 14, *Bolivina* cf. *B. danvillensis*, × 750, sutures and pores; Bahía San Blas

Fig. 15, *Bolivina* cf. *B. danvillensis*, × 650, aperture; Bahía San Blas

Fig. 16, *Bolivina difformis*, × 90; Golfo San Jorge

Fig. 17, *Bolivina difformis*, × 500, aperture; Golfo San Jorge

Fig. 18, *Bolivina doniezi*, × 240; Puerto Deseado

Fig. 19, *Bolivina doniezi*, × 250; Puerto Deseado

Fig. 20, *Bolivina doniezi*, × 1000, aperture; Puerto Deseado

Plate 3

Fig. 1, *Bolivina ordinaria*, × 150; Golfo San Jorge

Fig. 2, *Bolivina ordinaria*, × 150; Golfo San Jorge

Fig. 3, *Bolivina ordinaria*, × 1000, aperture; Golfo San Jorge

Fig. 4, *Bolivina pseudoplicata*, × 125; Puerto Deseado

Fig. 5, *Bolivina pseudoplicata*, × 125; Santa Cruz

Fig. 6, *Bolivina pseudoplicata*, × 225, apertural view; Santa Cruz

Fig. 7, *Bolivina pseudoplicata*, × 400, aperture; Puerto Deseado

Fig. 8, *Bolivina pseudoplicata*, × 300, wall surface showing the raised polygonal ridges; Puerto Deseado

Fig. 9, *Bolivina striatula*, × 125; Golfo San Jorge

Fig. 10, *Bolivina striatula*, × 100; La Paloma

Fig. 11, *Bolivina striatula*, × 150; Golfo San Jorge

Fig. 12, *Bolivina striatula*, × 125; Golfo San Jorge

Fig. 13, *Bolivina striatula*, × 500, aperture; La Paloma

Fig. 14, *Bolivina tortuosa*, × 175; Uruguayan coast

Fig. 15, *Bolivina tortuosa*, × 250, apertural view; Uruguayan coast

Fig. 16, *Bolivina tortuosa*, × 500, aperture; Uruguayan coast

Fig. 17, *Bolivina tortuosa*, × 200, apical end; Uruguayan coast

Fig. 18, *Bolivina translucens*, × 200; Puerto Deseado

Fig. 19, *Bolivina translucens*, × 190; Puerto Deseado

Fig. 20, *Bolivina translucens*, × 600, apertural view; Golfo San Jorge

Fig. 21, *Bolivina translucens*, × 700, aperture; Golfo San Jorge

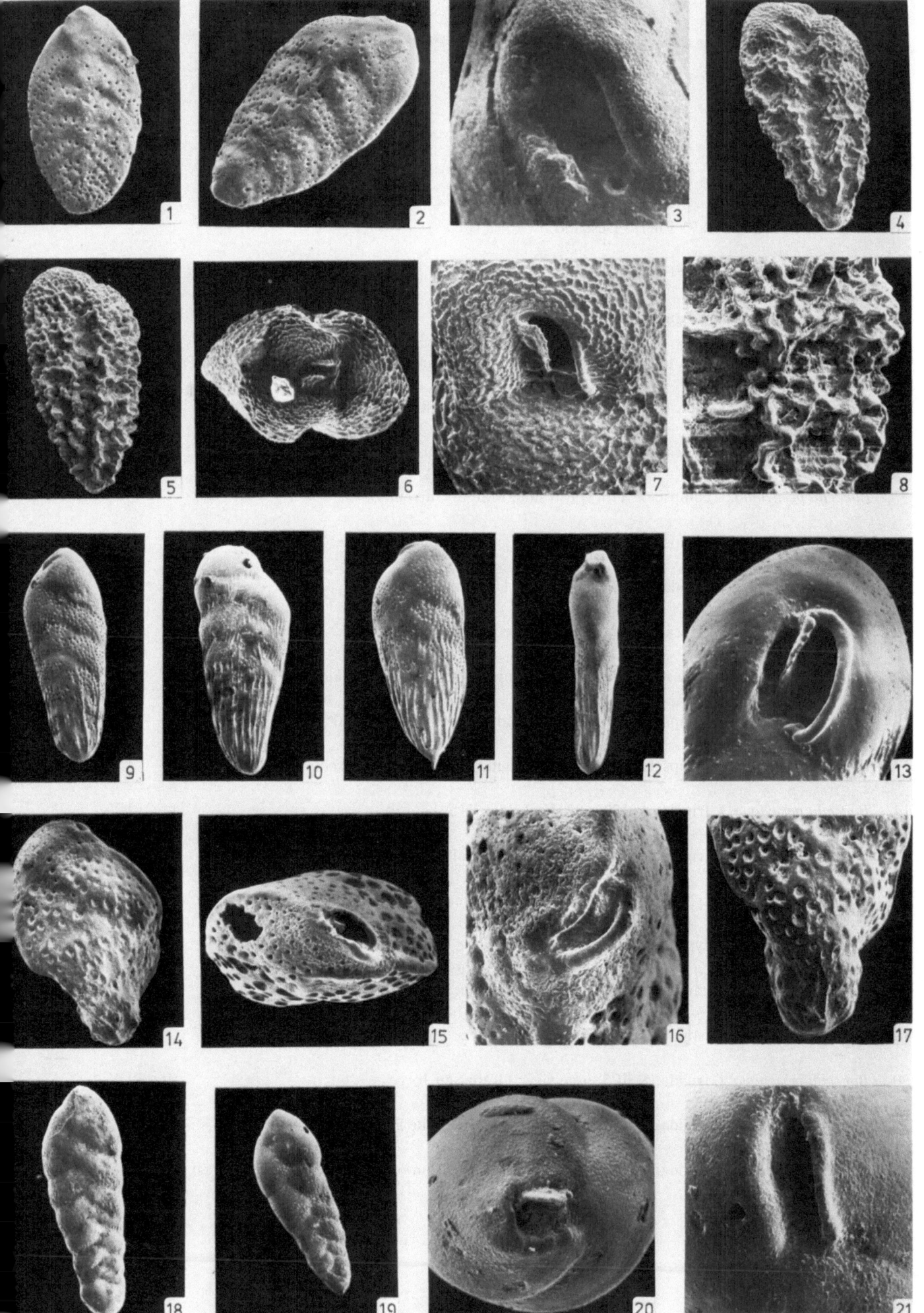

65

Plate 4

Fig. 1, *Bolivina variabilis*, × 150; Puerto Deseado

Fig. 2, *Bolivina variabilis*, × 125; Puerto Deseado

Fig. 3, *Bolivina variabilis*, × 500, aperture; Puerto Deseado

Fig. 4, *Bolivina variabilis*, × 500, test surface showing polygonal rims raised above pores; Puerto Deseado

Fig. 5, *Buccella peruviana*, f. typica, × 80, spiral side; Chilean coast

Fig. 6, *Buccella peruviana*, f. typica transitional to f. campsi, × 100, spiral side; Malvin Current, 51°30′S

Fig. 7, *Buccella peruviana*, f. campsi, × 110, spiral side; Malvin Current, 38°S

Fig. 8, *Buccella peruviana*, f. campsi transitional to f. frigida, × 100, spiral side; Tierra del Fuego

Fig. 9, *Buccella peruviana*, f. frigida, × 200, spiral side; Ushuaia

Fig. 10, *Buccella peruviana*, f. typica, × 80, umbilical side; Chilean coast

Fig. 11, *Buccella peruviana*, f. typica transitional to f. campsi, × 90, umbilical side; Malvin Current, 52°S

Fig. 12, *Buccella peruviana*, f. campsi, × 110, umbilical side; Malvin Current, 50°30′S

Fig. 13, *Buccella peruviana*, f. campsi transitional to f. frigida, × 100, umbilical side; Ushuaia

Fig. 14, *Buccella peruviana*, f. frigida, × 150, umbilical side; 37°S

Fig. 15, *Buccella peruviana*, f. frigida, × 160, umbilical side; Golfo San Jorge

Fig. 16, *Buccella peruviana*, f. typica, × 85, edge view; Chilean coast

Fig. 17, *Buccella peruviana*, f. typica transitional to f. campsi, × 125, edge view; Golfo San Jorge

Fig. 18, *Buccella peruviana*, f. campsi, × 100, edge view; Malvin Current, 38°S

Fig. 19, *Buccella peruviana*, f. campsi transitional to f. frigida, × 150, edge view; Puerto Deseado

Fig. 20, *Buccella peruviana*, f. frigida, × 175, edge view; Golfo San Jorge

Fig. 21, *Buccella peruviana*, f. frigida, × 150, edge view; estuary of the Río de la Plata

Fig. 22, *Buccella peruviana*, f. frigida, × 500, detail of the granulation of the umbilical side sutures; Ushuaia

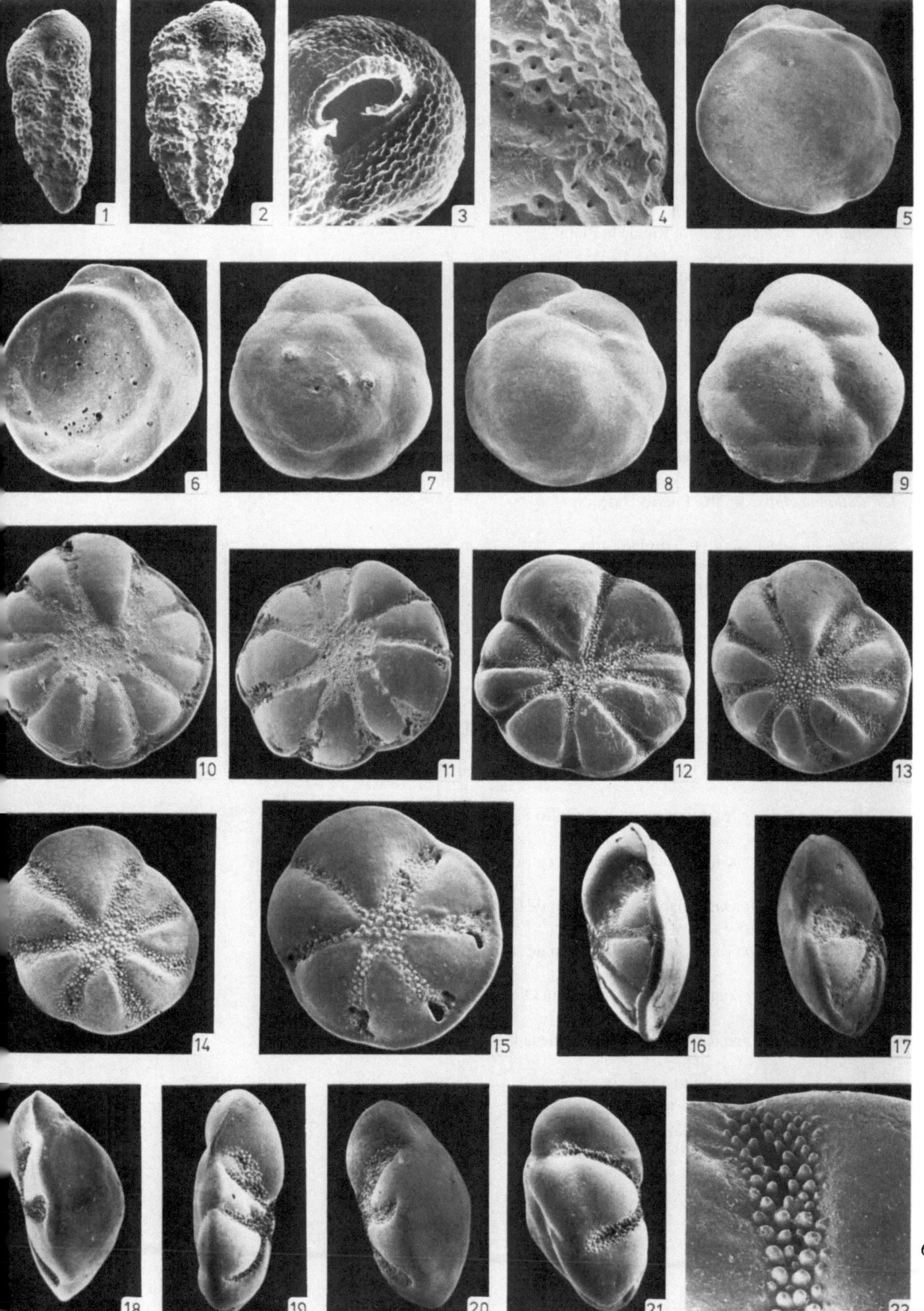

67

Plate 5

Fig. 1, *Bulimina aculeata*, × 140; Río de la Plata

Fig. 2, *Bulimina aculeata*, × 100; estuary of the Río de la Plata

Fig. 3, *Bulimina aculeata*, × 100; estuary of the Río de la Plata

Fig. 4, *Bulimina elongata*, × 100; Golfo San Jorge

Fig. 5, *Bulimina elongata*, × 100; Golfo San Jorge

Fig. 6, *Bulimina elongata*, × 400, aperture; 36°S, 53°W (outer shelf)

Fig. 7, *Bulimina gibba*, × 210; Puerto Deseado

Fig. 8, *Bulimina gibba*, × 70; Puerto Deseado

Fig. 9, *Bulimina gibba*, × 250, aperture; Puerto Deseado

Fig. 10, *Bulimina marginata*, × 110; Uruguayan coast

Fig. 11, *Bulimina marginata*, × 125; Golfo Nuevo

Fig. 12, *Bulimina marginata*, × 500, spines lining the base of a chamber; Golfo Nuevo

Fig. 13, *Bulimina patagonica*, f. typica, × 125; Golfo San Jorge

Fig. 14, *Bulimina patagonica*, f. typica, × 165; Golfo San Matías

Fig. 15, *Bulimina patagonica*, f. typica, × 140, apertural view; 37°S (inner shelf)

Fig. 16, *Bulimina patagonica*, f. glabra, × 145; Río de la Plata

Fig. 17, *Bulimina patagonica*, f. glabra, × 210; Río de la Plata

Fig. 18, *Bulimina* cf. *B. pseudoaffinis*, × 125; Caleta Olivia

Fig. 19, *Bulimina* cf. *B. pseudoaffinis*, × 215; Río de la Plata

Fig. 20, *Bulimina pupoides*, × 140; Golfo San Jorge

Fig. 21, *Bulimina pupoides*, × 175; Golfo San Jorge

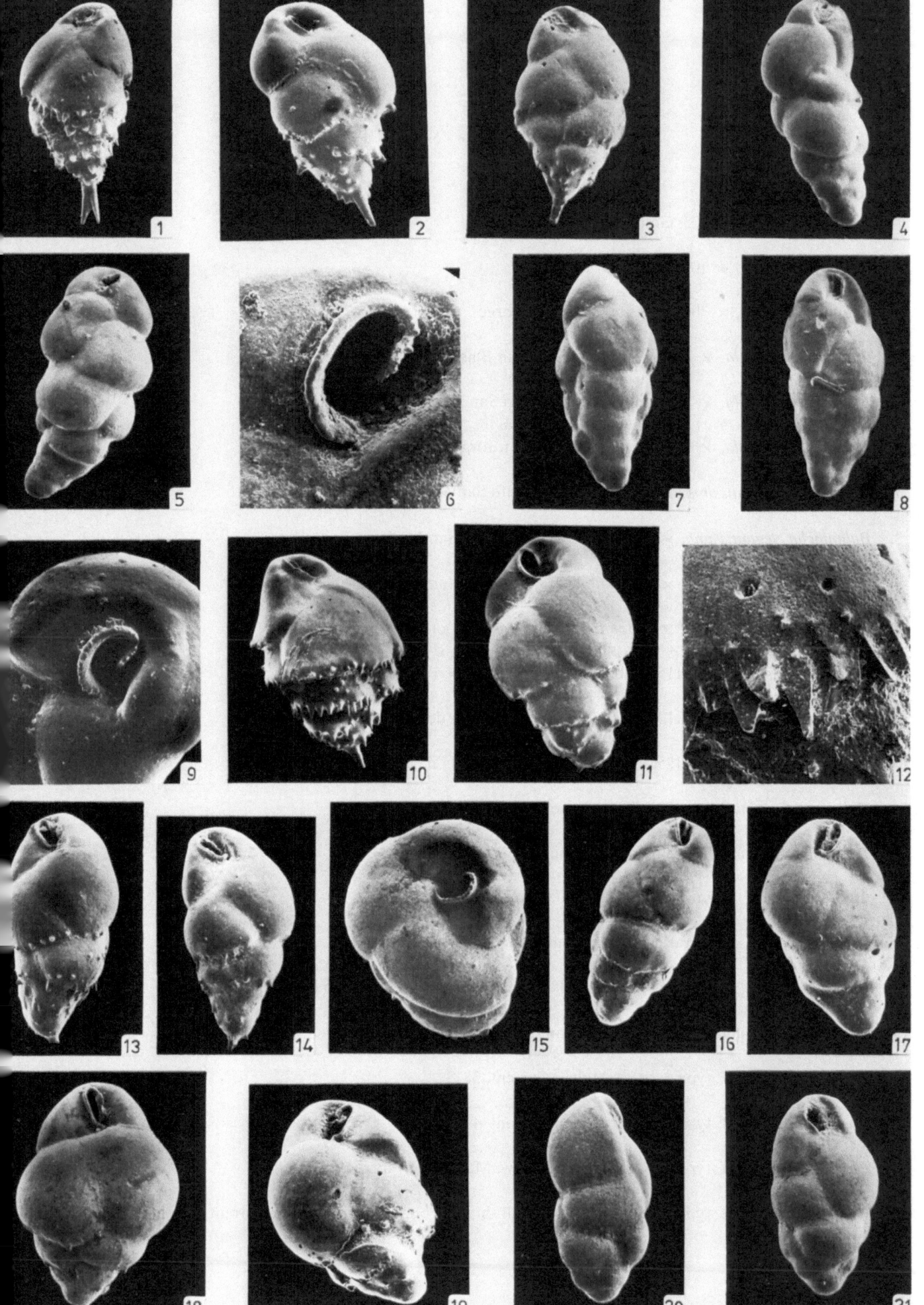

Plate 6

Fig. 1,   *Bulimina subulata*, × 105; Golfo San Jorge

Fig. 2,   *Bulimina subulata*, × 100; Golfo San Jorge

Fig. 3,   *Bulimina subulata*, × 300, aperture; Golfo San Jorge

Fig. 4,   *Buliminella auricula*, × 250, spiral side; Bahía San Blas

Fig. 5,   *Buliminella auricula*, × 250, umbilical side; Bahía San Blas

Fig. 6,   *Buliminella auricula*, × 275, edge view; Golfo San Jorge

Fig. 7,   *Buliminella elegantissima*, × 175, spiral side; Golfo San Matías

Fig. 8,   *Buliminella elegantissima*, × 175, umbilical side; Golfo San Matías

Fig. 9,   *Buliminella elegantissima*, × 180, umbilical side; Golfo San Matías

Fig. 10, *Buliminella elegantissima*, × 150, umbilical side; estuary of the Río de la Plata

Fig. 11, *Buliminella seminuda*, × 115, umbilical side; south of Tierra del Fuego

Fig. 12, *Buliminella seminuda*, × 110, spiral side; south of Tierra del Fuego

Fig. 13, *Buliminella seminuda*, × 125, umbilical side; north of Isla de las Estados

Fig. 14, *Buliminella seminuda*, × 100, edge view; south of Tierra del Fuego

Fig. 15, *Buliminella seminuda*, × 250, aperture; south of Tierra del Fuego

Fig. 16, *Cancris sagra*, × 75, umbilical side; Uruguayan coast

Fig. 17, *Cancris sagra*, × 50, spiral side; Uruguayan coast

Fig. 18, *Cancris sagra*, × 250, aperture; inner shelf off the mouth of the Río de la Plata

Fig. 19, *Cassidulina crassa*, f. typica, × 225; Malvin Current, 40°S, 97m

Fig. 20, *Cassidulina crassa*, f. typica, × 70; Malvin Current, 38°S

Fig. 21, *Cassidulina crassa*, f. typica, × 65; Malvin Current, 41°S, 78m

Fig. 22, *Cassidulina crassa*, f. typica, × 80, apertural view; Malvin Current, 44°S

Fig. 23, *Cassidulina crassa*, f. typica, × 400, detail of wall showing pore density; Malvin Current near Islas Malvinas

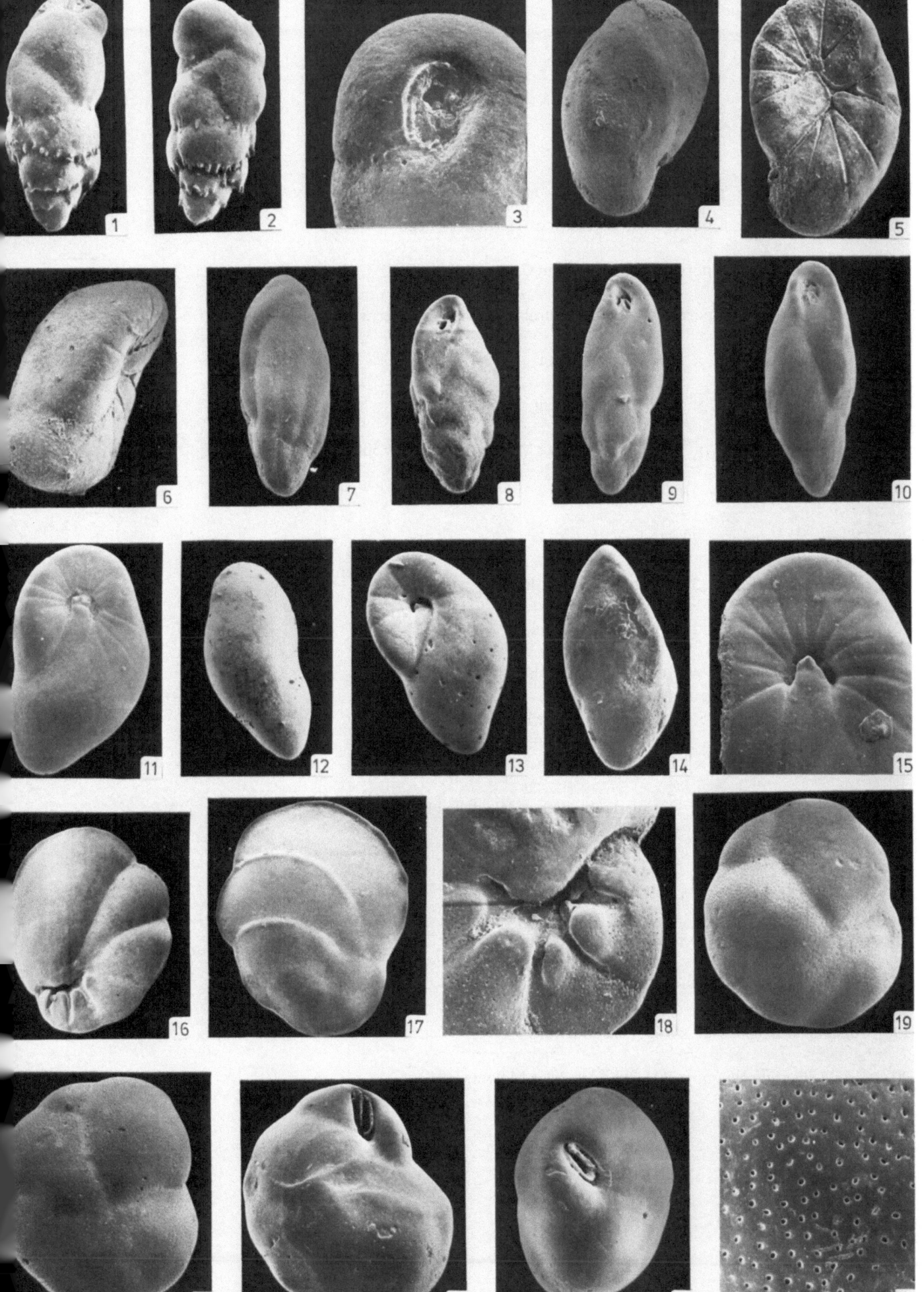

71

Plate 7

Fig. 1, *Cassidulina crassa*, f. porrecta, × 60; south of Tierra del Fuego

Fig. 2, *Cassidulina crassa*, f. porrecta, × 45; Malvin Current, 39°45′S

Fig. 3, *Cassidulina crassa*, f. porrecta, × 55, apertural view; south of Tierra del Fuego, 88-115m

Fig. 4, *Cassidulina laevigata*, × 250; Río de la Plata

Fig. 5, *Cassidulina laevigata*, × 300; Malvin Current, 37°50′S

Fig. 6, *Cassidulina laevigata*, × 225, apertural view; Malvin Current, 37°50′S

Fig. 7, *Cassidulina minuta*, × 210; Ushuaia

Fig. 8, *Cassidulina minuta*, × 200; Ushuaia

Fig. 9, *Cassidulina minuta*, × 300; Malvin Current, 46°S

Fig. 10, *Cassidulina minuta*, × 250, edge view; Ushuaia

Fig. 11, *Cassidulina minuta*, × 400, detail of wall; Ushuaia

Fig. 12, *Cassidulina pulchella*, × 115; north of Isla de los Estados

Fig. 13, *Cassidulina pulchella*, × 110; Tierra del Fuego

Fig. 14, *Cassidulina pulchella*, × 300, aperture; north of Isla de los Estados

Fig. 15, *Cassidulina rossensis*, × 115; Mar del Plata

Fig. 16, *Cassidulina rossensis*, × 190; Tierra del Fuego

Fig. 17, *Cassidulina rossensis*, × 400, detail of wall; Mar del Plata

Fig. 18, *Cassidulina subglobosa*, × 210; Islas Malvinas

Fig. 19, *Cassidulina subglobosa*, × 145; Islas Malvinas

Fig. 20, *Cassidulina subglobosa*, × 300, aperture; Islas Malvinas

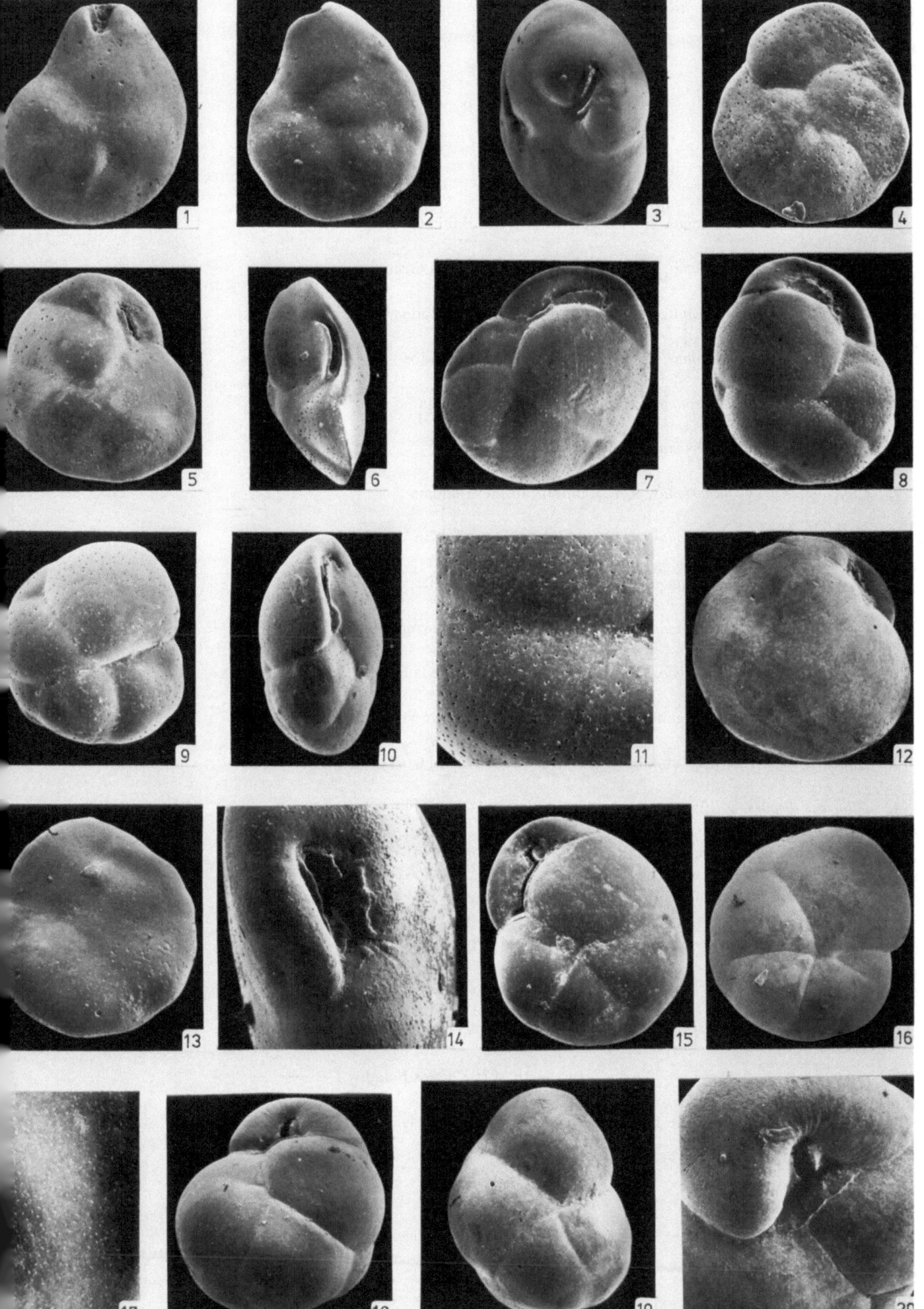

73

Plate 8

Fig. 1, *Cassidulinoides parkerianus*, × 115; Malvin Current, Bahía Grande

Fig. 2, *Cassidulinoides parkerianus*, × 100; Malvin Current, Bahía Grande

Fig. 3, *Cassidulinoides parkerianus*, × 100; north of Isla de los Estados

Fig. 4, *Cassidulinoides parkerianus*, × 85; north of Isla de los Estados

Fig. 5, *Cibicides* ex gr. *C. aknerianus*, × 90, umbilical side; Ushuaia

Fig. 6, *Cibicides* ex gr. *C. aknerianus*, × 90, spiral side; Ushuaia

Fig. 7, *Cibicides* ex gr. *C. aknerianus*, × 125, umbilical side; Ushuaia

Fig. 8, *Cibicides* ex gr. *C. aknerianus*, × 125, spiral side; Malvin Current, 51°S

Fig. 9, *Cibicides* ex gr. *C. aknerianus*, × 90, edge view; south of Tierra del Fuego

Fig. 10, *Cibicides* ex gr. *C. aknerianus*, × 50, spiral side; Malvin Current, 49°S

Fig. 11, *Cibicides* ex gr. *C. aknerianus*, × 75, umbilical side; Río de la Plata

Fig. 12, *Cibicides dispars*, × 100, umbilical side, without knob; Islas Malvinas (topotype)

Fig. 13, *Cibicides dispars*, × 150, umbilical side, with knob; Malvin Current, 43°S

Fig. 14, *Cibicides dispars*, × 125, umbilical side, with knob; Islas Malvinas (topotype)

Fig. 15, *Cibicides dispars*, × 85, spiral side; Islas Malvinas (topotype)

Fig. 16, *Cibicides dispars*, × 140, edge view; Islas Malvinas (topotype)

Fig. 17, *Cibicides* cf. *C. fletcheri*, × 100, umbilical side; Malvin Current, 49°10′S

Fig. 18, *Cibicides* cf. *C. fletcheri*, × 110, umbilical side; Malvin Current, 49°10′S

Fig. 19, *Cibicides* cf. *C. fletcheri*, × 125, spiral side; Malvin Current, 49°10′S

Fig. 20, *Cibicides* cf. *C. fletcheri*, × 90, edge view; north of Isla de los Estados

Fig. 21, *Cibicides* cf. *C. fletcheri*, × 125, edge view; Malvin Current, 49°10′S

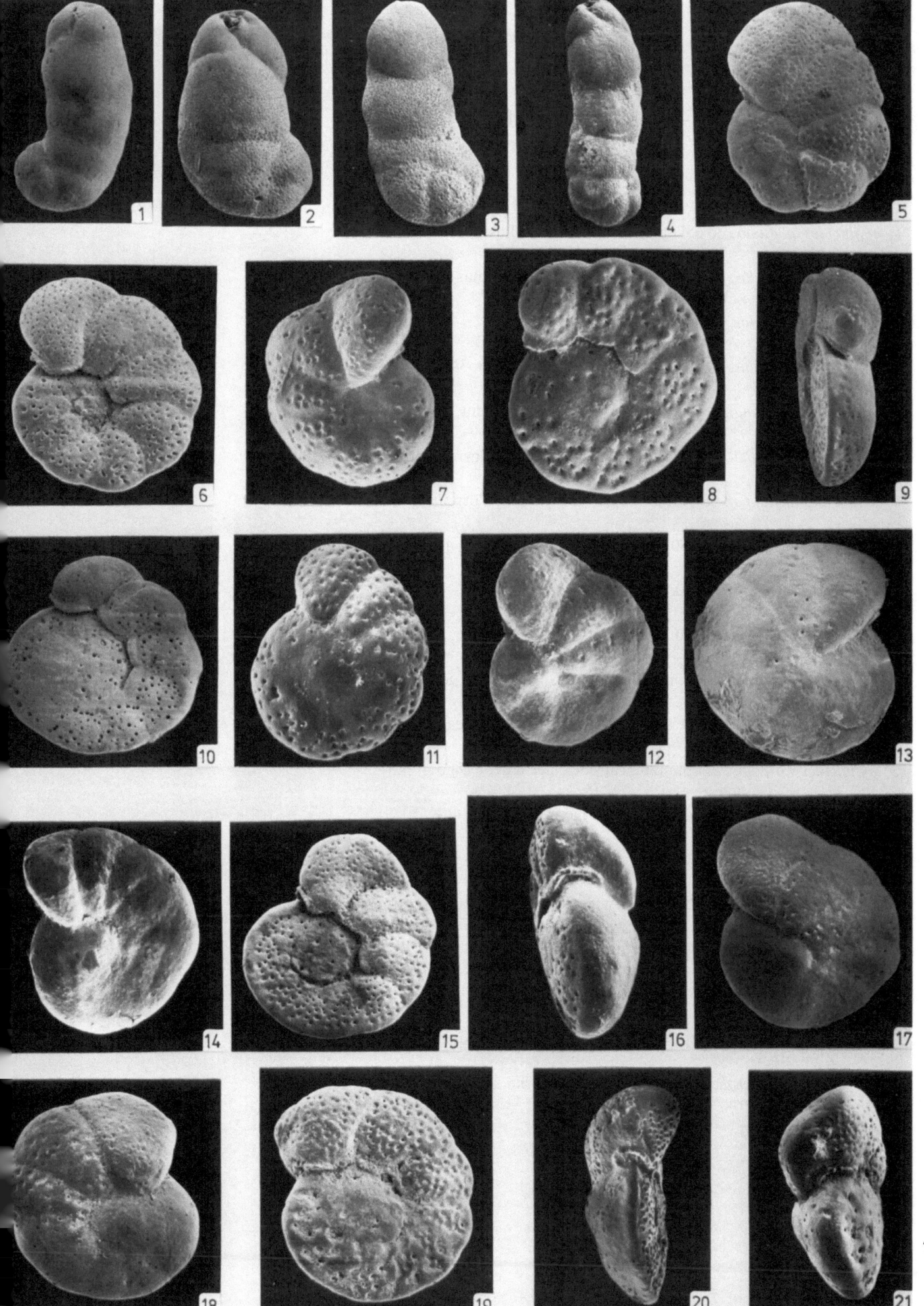

75

Plate 9

Fig. 1, *Cibicides lobatulus*, × 65, umbilical side; Islas Malvinas

Fig. 2, *Cibicides lobatulus*, × 90, umbilical side; Islas Malvinas

Fig. 3, *Cibicides lobatulus*, × 90, spiral side; Islas Malvinas

Fig. 4, *Cibicides lobatulus*, × 80, edge view; Malvin Current, 34°S

Fig. 5, *Cibicides mckannai*, × 175, spiral side; Malvin Current, 45°S

Fig. 6, *Cibicides mckannai*, × 225, umbilical side; Malvin Current, 46°S

Fig. 7, *Cibicides mckannai*, × 175, umbilical side; Malvin Current, 39°30′S

Fig. 8, *Cibicides mckannai*, × 250, edge view; Malvin Current, 45°S

Fig. 9, *Cibicides refulgens*, × 100, edge view; Golfo San Jorge

Fig. 10, *Cibicides refulgens*, × 100, spiral side; Golfo San Jorge

Fig. 11, *Cibicides refulgens*, × 90, umbilical side; Golfo San Jorge

Fig. 12, *Cibicides variabilis*, × 60, umbilical side; Tierra del Fuego

Fig. 13, *Cibicides variabilis*, × 75, umbilical side; Tierra del Fuego

Fig. 14, *Cibicides variabilis*, × 60, spiral side; Tierra del Fuego

Fig. 15, *Cibicides variabilis*, × 80, spiral side; Tierra del Fuego

Fig. 16, *Cibicides variabilis*, × 75, edge view; north of Isla de las Estados

Fig. 17, *Cibicides variabilis*, × 200, aperture; Tierra del Fuego

Fig. 18, *Cribrorotalia meridionalis*, × 60, spiral side; Ushuaia

Fig. 19, *Cribrorotalia meridionalis*, × 85, umbilical side; Ushuaia

Fig. 20, *Cribrorotalia meridionalis*, × 85, edge view; Ushuaia

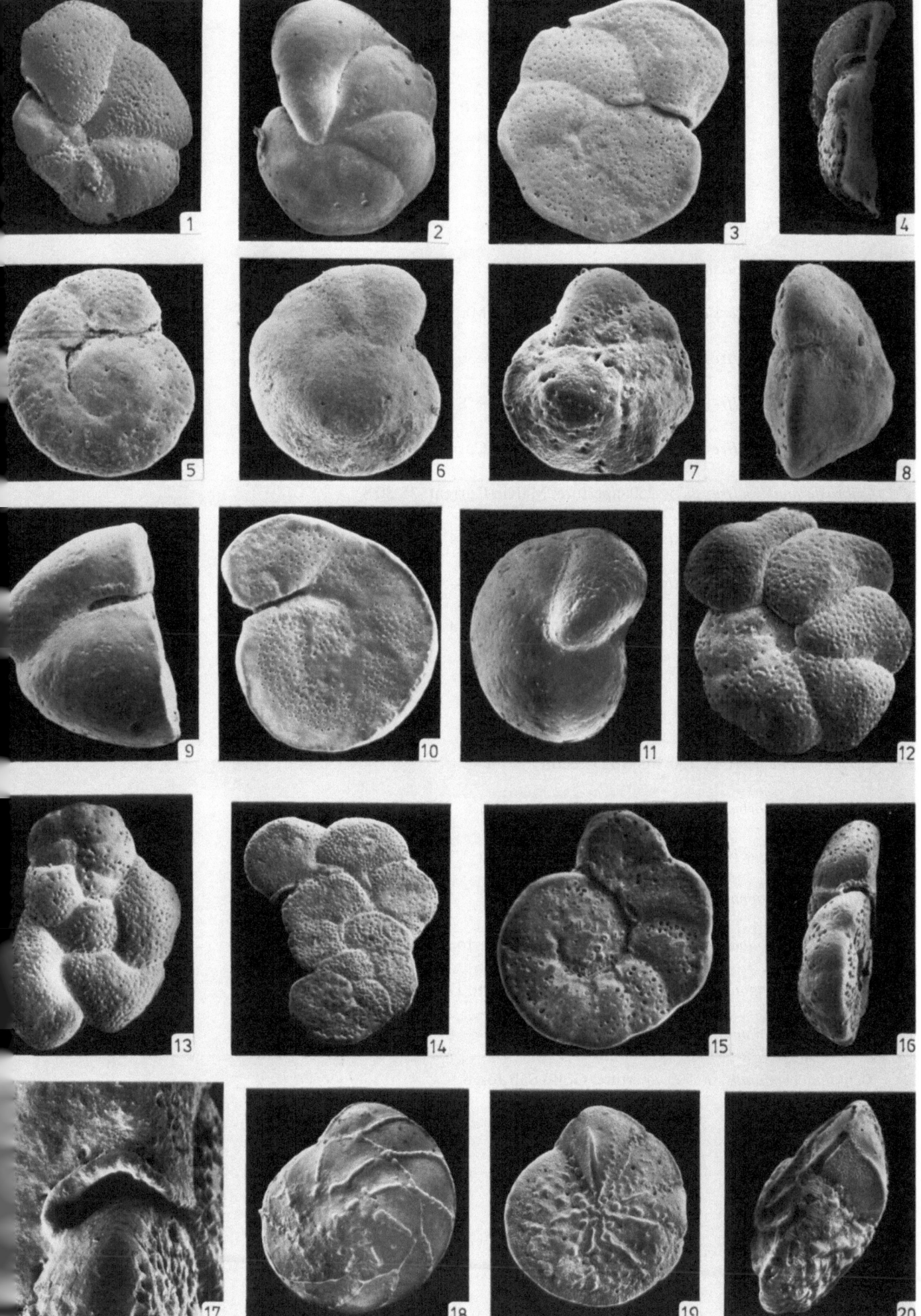

Plate 10

Fig. 1, *Cribrostomoides crassimargo*, × 40; Malvin Current, 36°S

Fig. 2, *Cribrostomoides crassimargo*, × 40, edge view; Malvin Current, 36°S

Fig. 3, *Cribrostomoides crassimargo*, × 150, detail of wall; Malvin Current, 36°S

Fig. 4, *Cribrostomoides jeffreysii*, × 60; Malvin Current, 35°30′S

Fig. 5, *Cribrostomoides jeffreysii*, × 200; Malvin Current, 48°S

Fig. 6, *Cribrostomoides jeffreysii*, × 65, edge view; Malvin Current, 35°30′S

Fig. 7, *Cribrostomoides jeffreysii*, × 250, aperture; Malvin Current, 35°30′S

Fig. 8, *Cribrostomoides weddellensis*, × 150; Malvin Current, 44°50′S

Fig. 9, *Cribrostomoides weddellensis*, × 210; Malvin Current, 44°S

Fig. 10, *Cribrostomoides weddellensis*, × 150; Malvin Current, 44°S

Fig. 11, *Cyclogyra involvens*, × 100; Puerto Deseado

Fig. 12, *Cyclogyra involvens*, × 125, edge view; Puerto Deseado

Fig. 13, *Cyclogyra planorbis*, × 150; Puerto Deseado

Fig. 14, *Cyclogyra planorbis*, × 200, edge view; Río Quequén

Fig. 15, *Cyclogyra planorbis*, × 200, edge view; Río Quequén

Fig. 16, *Dahlgrenia patagoniensis*, × 140; Puerto Deseado

Fig. 17, *Dahlgrenia patagoniensis*, × 175, apertural view; Puerto Deseado

Fig. 18, *Dahlgrenia patagoniensis*, × 750, detail of wall; Puerto Deseado

Fig. 19, *Dentalina communis*, × 25; Malvin Current, 34°30′S

Fig. 20, *Dentalina communis*, × 500, aperture; Golfo San Jorge

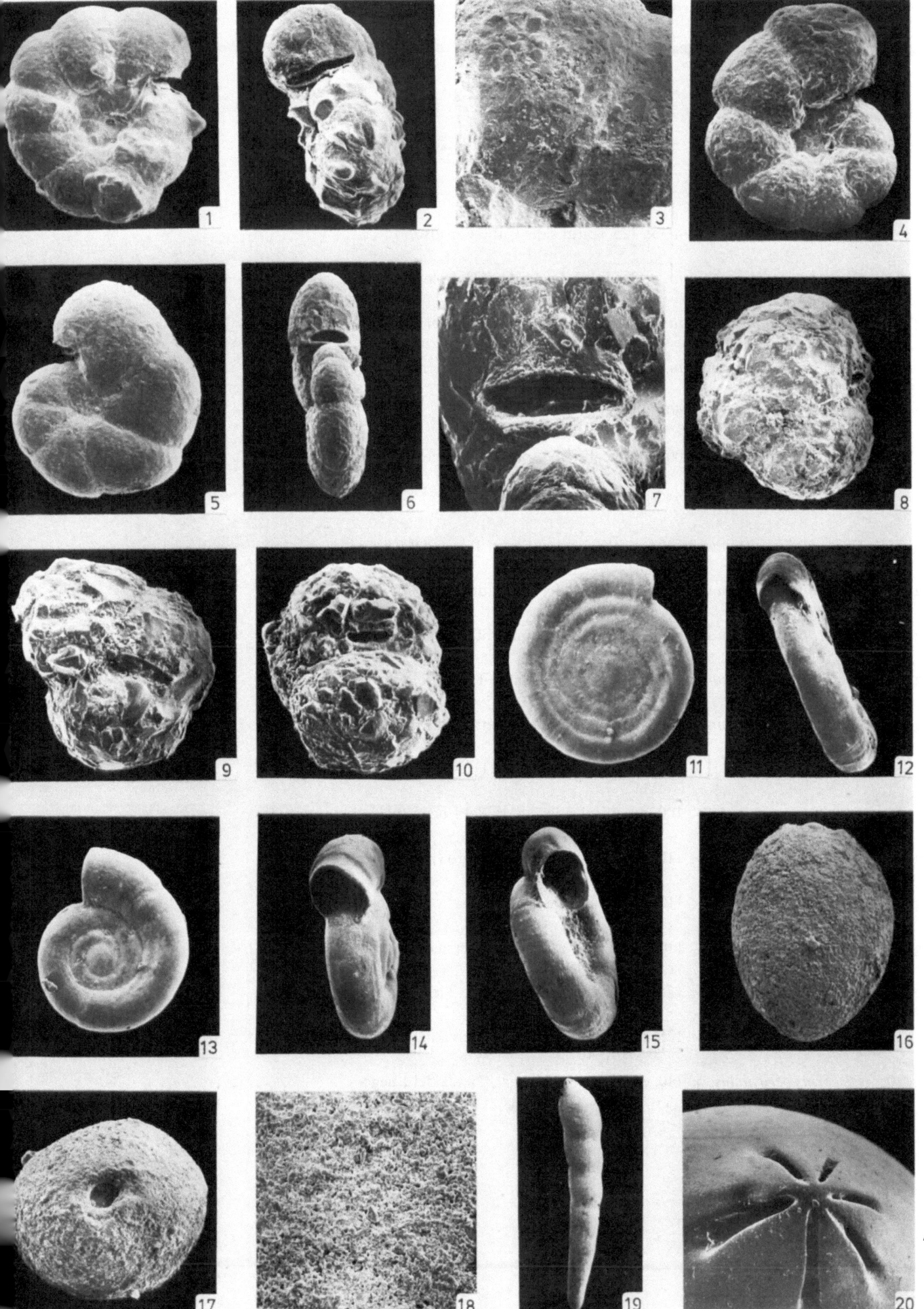

Plate 11

Fig. 1,   *Discorbinella altocamerata*, × 125, spiral side; Cabo Curioso

Fig. 2,   *Discorbinella altocamerata*, × 225, umbilical side; Isla de los Estados

Fig. 3,   *Discorbinella altocamerata*, × 120, edge view; south of Tierra del Fuego

Fig. 4,   *Discorbinella altocamerata*, × 500, aperture; south of Tierra del Fuego

Fig. 5,   *Discorbis bertheloti*, × 150, spiral side; Golfo San Jorge

Fig. 6,   *Discorbis bertheloti*, × 200, umbilical side; Golfo San Jorge

Fig. 7,   *Discorbis bertheloti*, × 225, edge view; Golfo San Jorge

Fig. 8,   *Discorbis isabelleanus*, × 30, spiral side; east coast. Tierra del Fuego

Fig. 9,   *Discorbis isabelleanus*, × 30, umbilical side: Tierra del Fuego

Fig. 10, *Discorbis isabelleanus*, × 25, umbilical side; south of Tierra del Fuego

Fig. 11, *Discorbis isabelleanus*, × 30, edge view; east coast, Tierra del Fuego

Fig. 12, *Discorbis isabelleanus*, × 85, aperture; south of Tierra del Fuego

Fig. 13, *Discorbis malovensis*, × 100, spiral side; Tierra del Fuego (probably topotype)

Fig. 14, *Discorbis malovensis*, × 100, umbilical side; Tierra del Fuego (probably topotype)

Fig. 15, *Discorbis malovensis*, × 110, edge view; Terra del Fuego (probably topotype)

Fig. 16, *Discorbis peruvianus*, × 110, spiral side; east coast, Tierra del Fuego

Fig. 17, *Discorbis peruvianus*, × 110, spiral side; east coast, Tierra del Fuego

Fig. 18, *Discorbis peruvianus*, × 125, umbilical side; east coast, Tierra del Fuego

Fig. 19, *Discorbis peruvianus*, × 125, edge view; east coast, Tierra del Fuego

Fig. 20, *Discorbis peruvianus*, × 110, edge view; east coast, Tierra del Fuego

81

Plate 12

Fig. 1, *Discorbis* cf. *D. valvulatus*, × 140, spiral side; La Paloma

Fig. 2, *Discorbis* cf. *D. valvulatus*, × 160, umbilical side; La Paloma

Fig. 3, *Discorbis* cf. *D. valvulatus*, × 175, umbilical side; La Paloma

Fig. 4, *Discorbis* cf. *D. valvulatus*, × 150, edge view; La Paloma

Fig. 5, *Discorbis williamsoni*, s.l., × 200, spiral side; Río de la Plata

Fig. 6, *Discorbis williamsoni*, s.l., × 100, spiral side; Río de la Plata

Fig. 7, *Discorbis williamsoni*, s.l., × 90, umbilical side; south of Isla de los Estados

Fig. 8, *Discorbis williamsoni*, s.l., × 110, edge view; south of Isla de los Estados

Fig. 9, *Discorbis williamsoni*, s.l., × 240, umbilical view; Río de la Plata

Fig. 10, *Discorbis williamsoni*, s.l., × 100, umbilical view; Río de la Plata

Fig. 11, *Discorbis williamsoni*, s.l., × 160, umbilical view; Miramar

Fig. 12, *Discorbis williamsoni*, s.l., × 200, edge view; Río de la Plata

Fig. 13, *Ehrenbergina pupa*, × 95, edge view; Islas Malvinas (topotype)

Fig. 14, *Ehrenbergina pupa*, × 75, edge view; Islas Malvinas (topotype)

Fig. 15, *Ehrenbergina pupa*, × 75, lateral view; Islas Malvinas (topotype)

Fig. 16, *Elphidium advenum depressulum*, × 200; Bahía San Blas

Fig. 17, *Elphidium advenum depressulum*, × 250; Estuary of the Río de la Plata

Fig. 18, *Elphidium advenum depressulum*, × 200, edge view; Bahía San Blas

Fig. 19, *Elphidium alvarezianum*, × 100; Islas Malvinas (topotype?)

Fig. 20, *Elphidium alvarezianum*, × 100, edge view; Islas Malvinas (topotype?)

Fig. 21, *Elphidium alvarezianum*, × 400, aperture; Islas Malvinas (topotype?)

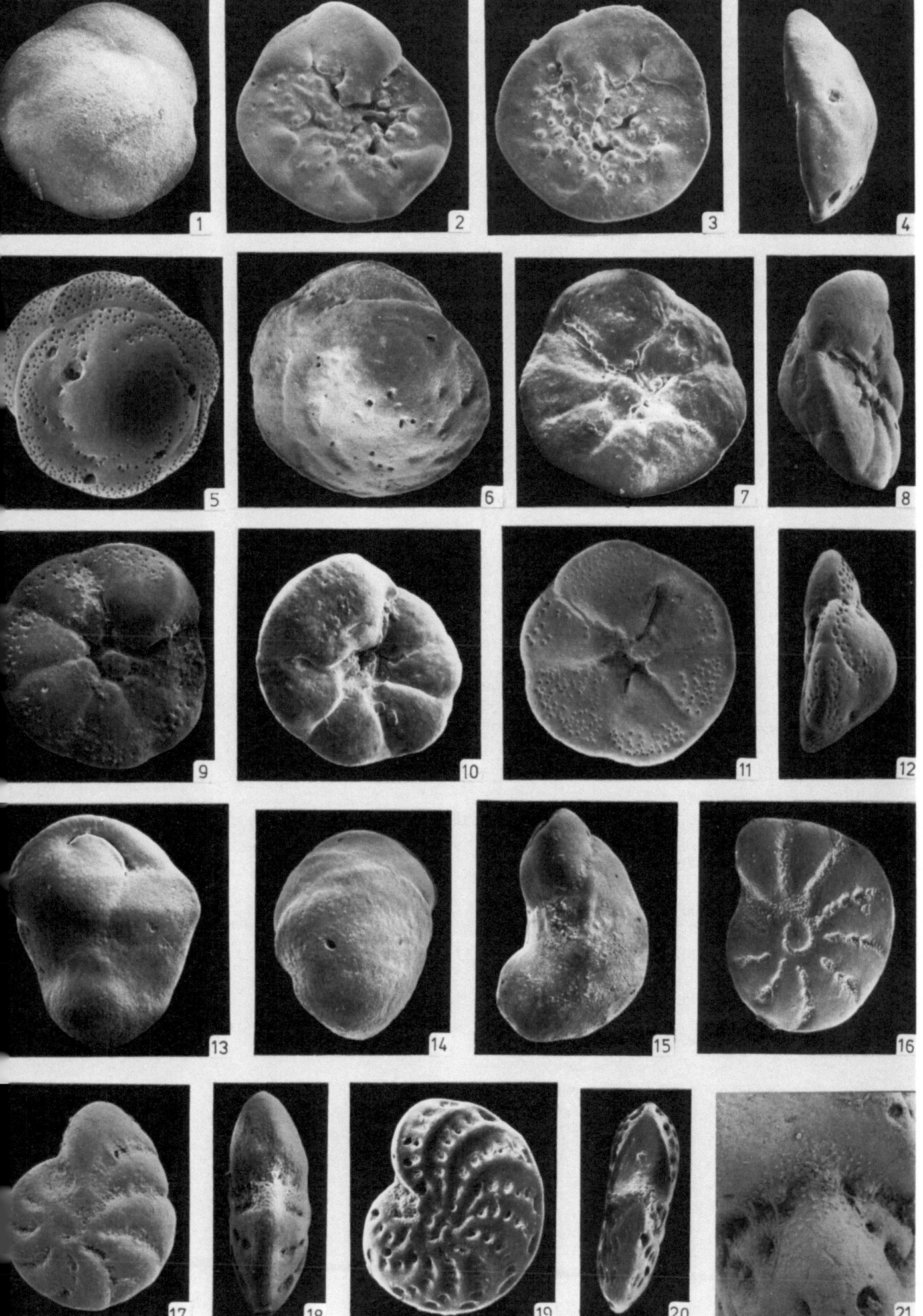

Plate 13

Fig. 1, *Elphidium articulatum*, × 95; Puerto Deseado

Fig. 2, *Elphidium articulatum*, × 110; Puerto Deseado

Fig. 3, *Elphidium articulatum*, × 100, edge view; Ushuaia

Fig. 4, *Elphidium articulatum*, × 250, aperture; Punta Tombo

Fig. 5, *Elphidium discoidale*, × 75; Río de la Plata

Fig. 6, *Elphidium discoidale*, × 65; central Uruguayan coast

Fig. 7, *Elphidium discoidale*, × 85, edge view; Río de la Plata

Fig. 8, *Elphidium excavatum*, × 150; Río Quequén

Fig. 9, *Elphidium excavatum*, × 225; Río Quequén

Fig. 10, *Elphidium excavatum*, × 200; Lagoa dos Patos

Fig. 11, *Elphidium excavatum*, × 250, edge view; Río Quequén

Fig. 12, *Elphidium galvestonense*, × 125; littoral zone, ≈ 37°S

Fig. 13, *Elphidium galvestonense*, × 125; littoral zone, ≈ 37°S

Fig. 14, *Elphidium galvestonense*, × 125, edge view; south Brazilian coast

Fig. 15, *Elphidium gunteri*, × 100; Lagoa dos Patos

Fig. 16, *Elphidium gunteri*, × 200; Bahía San Blas

Fig. 17, *Elphidium gunteri*, × 120, edge view; littoral zone, ≈ 37°S

Fig. 18, *Elphidium gunteri*, × 295, aperture; Bahía San Julián

Fig. 19, *Elphidium lessonii*, × 55; south of Tierra del Fuego

Fig. 20, *Elphidium lessonii*, × 75, edge view; north of Isla de los Estados

Plate 14

Fig. 1, *Elphidium macellum*, f. typica, × 60; Puerto Deseado

Fig. 2, *Elphidium macellum*, f. typica, × 70; Tierra del Fuego

Fig. 3, *Elphidium macellum*, f. typica, × 210, aperture; Tierra del Fuego

Fig. 5, *Elphidium macellum*, f. oweniana, × 125; north of Isla de los Estados

Fig. 6, *Elphidium macellum*, f. oweniana, × 100, edge view; north of Isla de los Estados

Fig. 6, *Elphidium macellum*, f. owenianum, × 100, edge view; north of Isla de los Estados

Fig. 7, *Elphidium magellanicum*, × 145; Puerto Deseado

Fig. 8, *Elphidium magellanicum*, × 175, edge view; Puerto Deseado

Fig. 9, *Elphidium magellanicum*, × 200, edge view; Puerto Deseado

Fig. 10, *Elphidium magellanicum*, × 400, aperture; Puerto Deseado

Fig. 11, *Elphidium margaritaceum*, × 185; Puerto Deseado

Fig. 12, *Elphidium margaritaceum*, × 150; Río de la Plata

Fig. 13, *Elphidium margaritaceum*, × 175, edge view; estuary of the Río de la Plata

Fig. 14, *Epistominella exigua*, × 300, spiral side; Malvin Current, 50°45′S

Fig. 15, *Epistominella exigua*, × 200, umbilical side; Río de la Plata

Fig. 16, *Epistominella exigua*, × 225, edge view; Río de la Plata

Fig. 17, *Epistominella exigua*, × 300, edge view; Río de la Plata

Fig. 18, *Fissurina auriculata*, × 150; Islas Malvinas

Fig. 19, *Fissurina auriculata*, × 210, edge view; Malvin Current, 49°10′S

Fig. 20, *Fissurina auriculata*, × 250, apertural view; Malvin Current, 49°10′S

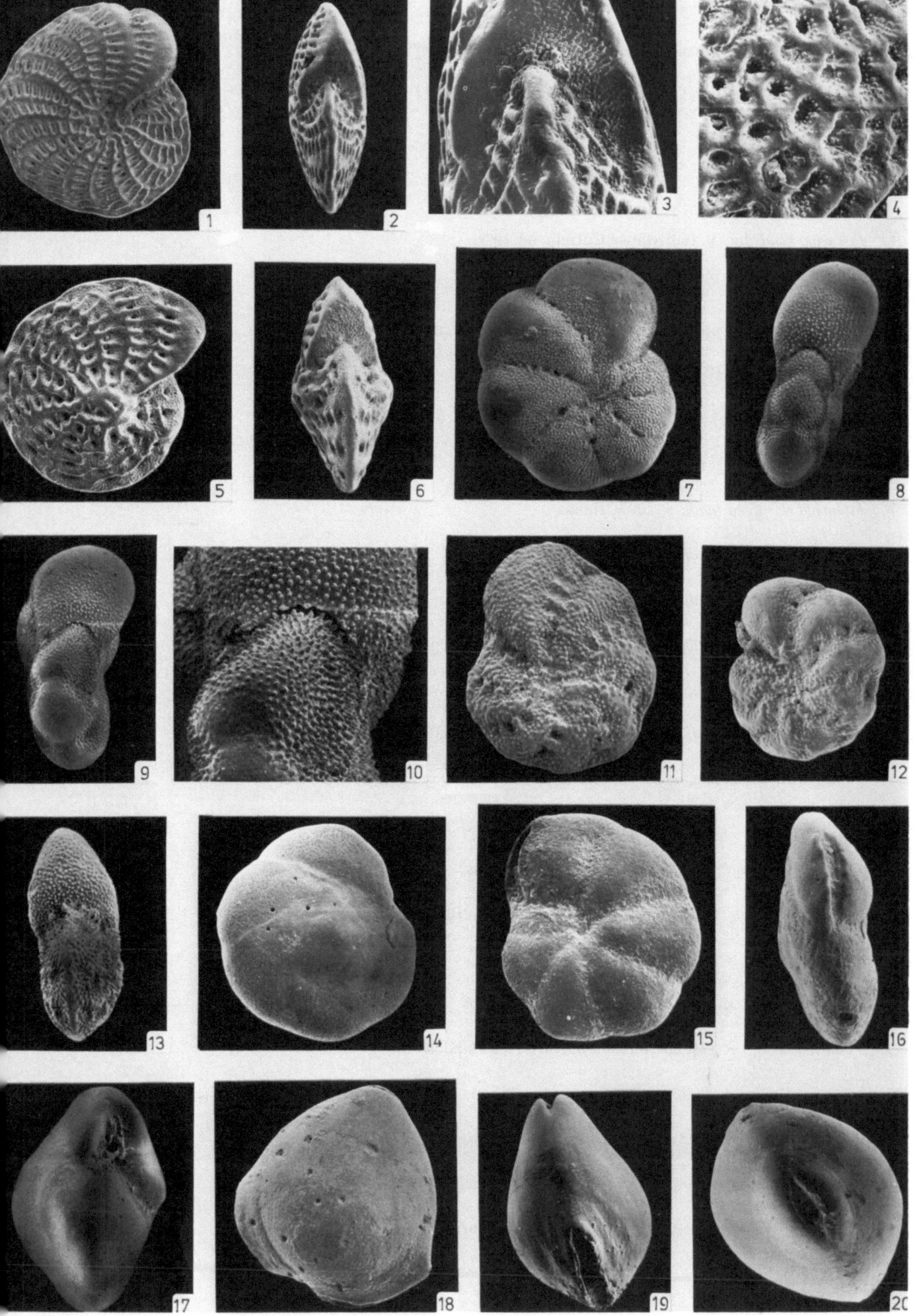

Plate 15

Fig. 1, *Fissurina bisulcata*, × 160; Malvin Current, 44°10'S

Fig. 2, *Fissurina bisulcata*, × 190, edge view; Malvin Current, 44°10'S

Fig. 3, *Fissurina bisulcata*, × 190, apertural view; Malvin Current, 44°10'S

Fig. 4, *Fissurina compressa*, × 200; Puerto Deseado

Fig. 5, *Fissurina compressa*, × 250, edge view; Puerto Deseado

Fig. 6, *Fissurina compressa*, × 300, apertural view; Ushuaia

Fig. 7, *Fissurina* aff. *F. earlandi*, × 200; Ushuaia

Fig. 8, *Fissurina* aff. *F. earlandi*, × 200, edge view; Ushuaia

Fig. 9, *Fissurina* aff. *F. earlandi*, × 250, apertural view; Ushuaia

Fig. 10, *Fissurina* aff. *F. earlandi*, × 1500, detail of wall showing pores and slits; Ushuaia

Fig. 11, *Fissurina* cf. *F. elliptica*, × 200; Puerto Deseado

Fig. 12, *Fissurina* cf. *F. elliptica*, × 200, edge view; Puerto Deseado

Fig. 13, *Fissurina* cf. *F. elliptica*, × 225, apertural view; Puerto Deseado

Fig. 14, *Fissurina laevigata*, × 200; Bahía San Blas

Fig. 15, *Fissurina laevigata*, × 200, edge view; Bahía San Blas

Fig. 16, *Fissurina laevigata*, × 250, apertural view; Bahía San Blas

Fig. 17, *Fissurina lucida*, × 200; Necochea

Fig. 18, *Fissurina lucida*, × 125, edge view; Isla de los Estados

Fig. 19, *Fissurina lucida*, × 175, apertural view; Isla de los Estados

Fig. 20, *Fissurina lucida*, × 2000, detail of opaque band (band is located in upper right); Necochea

Fig. 21, *Fissurina laureata*, × 300; Malvin Current, 50°40'S

Fig. 22, *Fissurina laureata*, × 300, edge view; Malvin Current, 50°40'S

Fig. 23, *Fissurina laureata*, × 350, apertural view; Malvin Current, 50°40'S

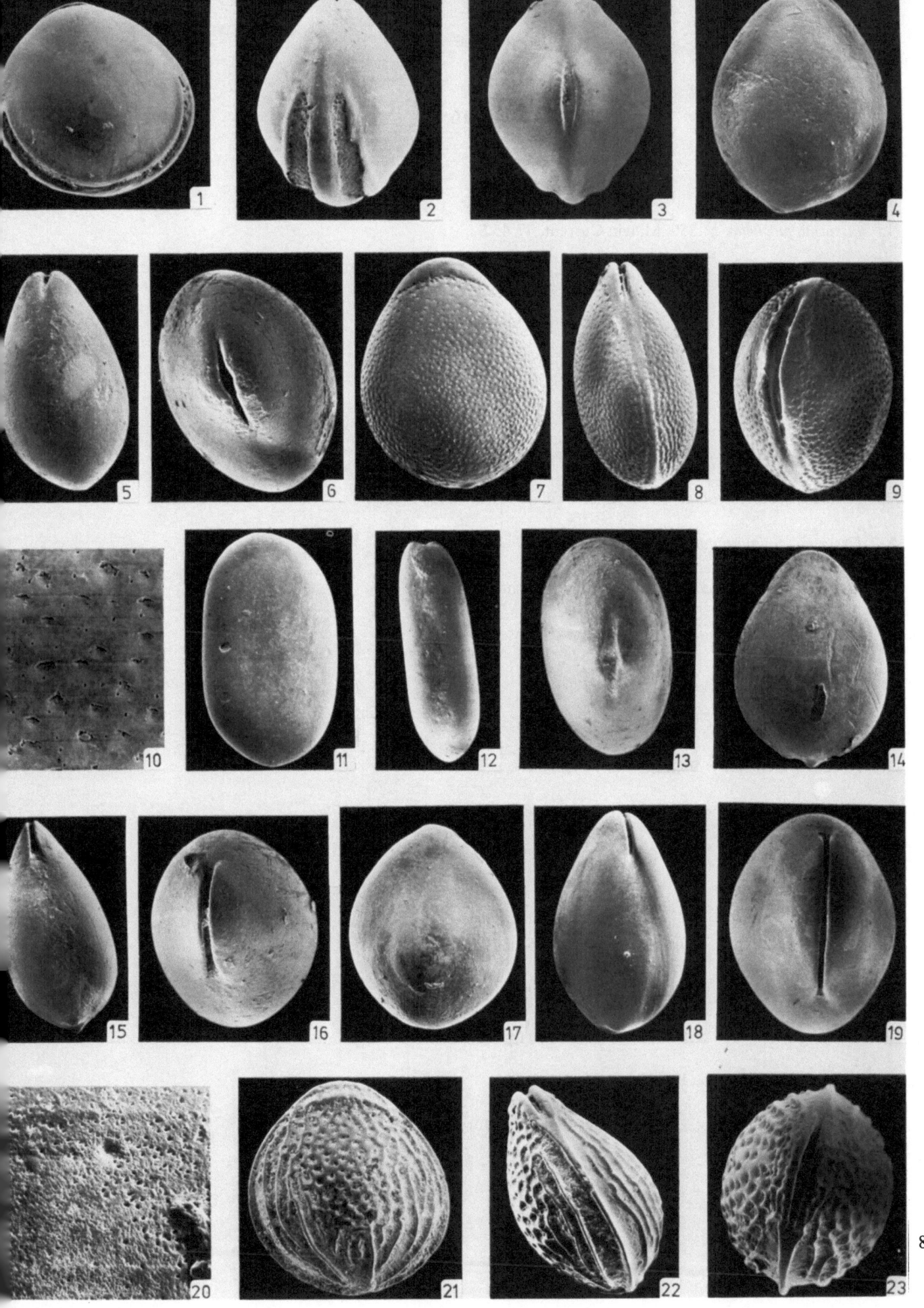

Plate 16

Fig. 1,   *Fissurina pulchella*, × 250; Malvin Current, 47°45′S

Fig. 2,   *Fissurina pulchella*, × 300; Malvin Current, 47°45′S

Fig. 3,   *Fissurina pulchella*, × 300, edge view; Malvin Current, 47°45′S

Fig. 4,   *Fissurina pulchella*, × 350, apertural view; Malvin Current, 47°45S

Fig. 5,   *Fissurina quadricostulata*, × 85; Tierra del Fuego

Fig. 6,   *Fissurina quadricostulata*, × 150, edge view; Estrecho de Magallenes

Fig. 7,   *Fissurina quadricostulata*, × 200, apertural view; Isla de los Estados

Fig. 8,   *Fissurina semimarginata*, × 150; Ushuaia

Fig. 9,   *Fissurina semimarginata*, × 175, edge view; Ushuaia

Fig. 10,   *Fissurina semimarginata*, × 250, apertural view; Ushuaia

Fig. 11,   *Florilus grateloupi*, × 90; Ushuaia

Fig. 12,   *Florilus grateloupi*, × 90; Ushuaia

Fig. 13,   *Florilus grateloupi*, × 90, edge view; Ushuaia

Fig. 14,   *Florilus grateloupi*, × 250, aperture; inner shelf, southern Brazil coast

Fig. 15,   *Florilus pauperatus*, × 225, Puerto Deseado

Fig. 16,   *Florilus pauperatus*, × 200; Puerto Deseado

Fig. 17,   *Florilus pauperatus*, × 300, edge view; Neocochea

Fig. 18,   *Florilus pauperatus*, × 300, edge view; Malvin Current, 44°50′S

Fig. 19,   *Florilus punctulatus*, × 60; estuary of the Río de la Plata

Fig. 20,   *Florilus punctulatus*, × 175; Golfo San Jorge

Fig. 21,   *Florilus punctulatus*, × 250, edge view; Golfo San Jorge

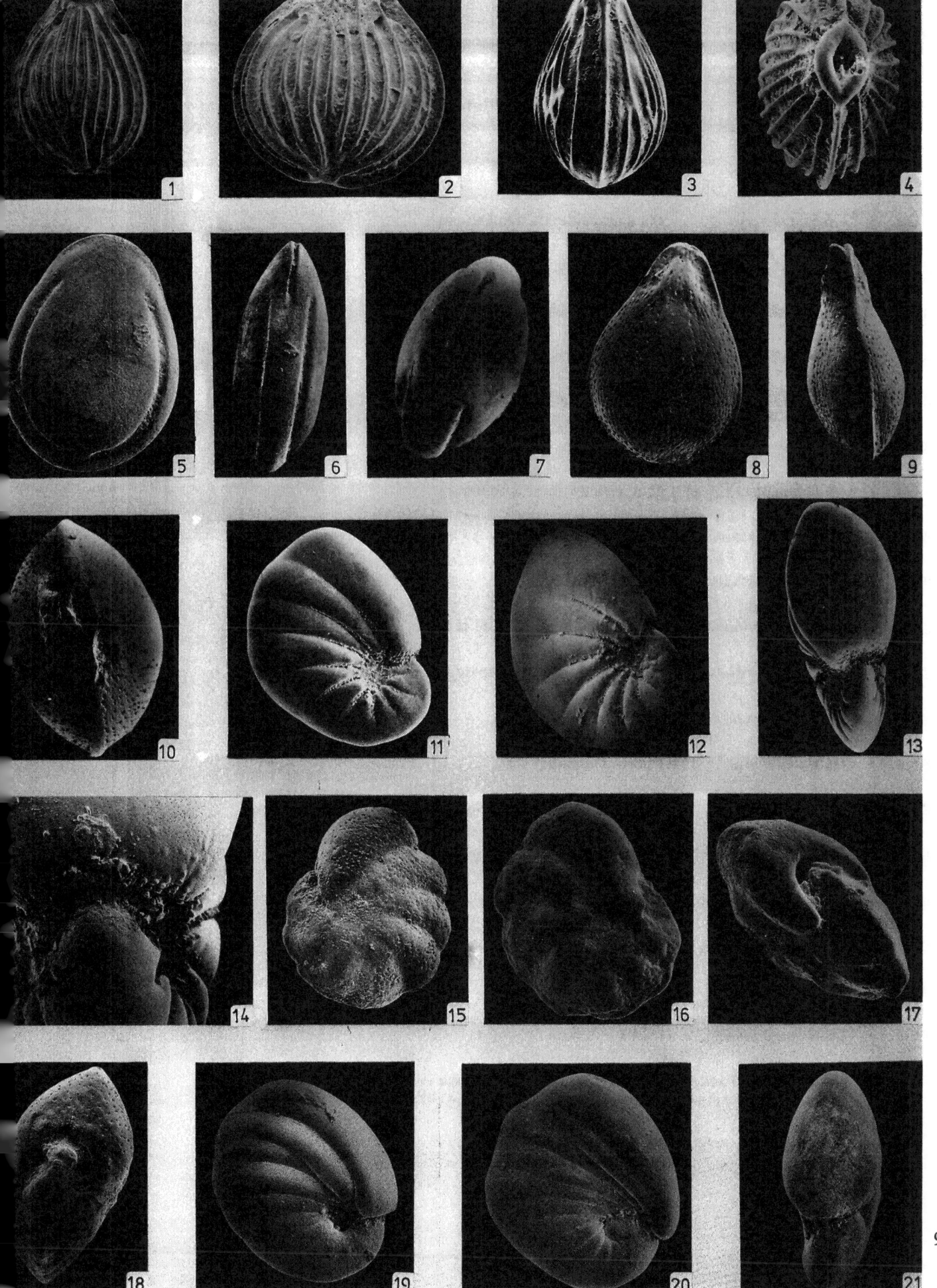

Plate 17

Fig. 1, *Glabratella chasteri*, × 500, spiral side; Puerto Deseado

Fig. 2, *Glabratella chasteri*, × 400, spiral side; Puerto Deseado

Fig. 3, *Glabratella chasteri*, × 550, edge view; Puerto Deseado

Fig. 4, *Glabratella chasteri*, × 450, umbilical side; Puerto Deseado

Fig. 5, *Globulina australis*, × 150; La Paloma

Fig. 6, *Globulina australis*, × 100; La Paloma

Fig. 7, *Globulina australis*, × 150, apertual view; La Paloma

Fig. 8, *Globulina caribaea*, × 75; Río de la Plata

Fig. 9, *Globulina caribaea*, × 100; Río de la Plata

Fig. 10, *Globulina caribaea*, × 100, apertural view; Río de la Plata

Fig. 11, *Globulina caribaea*, × 400, detail of the wall; Río de la Plata

Fig. 12, *Guttulina lactea*, × 90; Puerto Deseado

Fig. 13, *Guttulina lactea*, × 115; Puerto Deseado

Fig. 14, *Guttulina lactea*, × 150, apertural view; Puerto Deseado

Fig. 15, *Guttulina plancii*, × 75; Bahía Sanguineto

Fig. 16, *Guttulina plancii*, × 80; Bahía Sanguineto

Fig. 17, *Guttulina plancii*, × 150, apertural view; Bahía Sanguineto

Fig. 18, *Guttulina problema*, × 125; La Paloma

Fig. 19, *Guttulina problema*, × 135; La Paloma

Fig. 20, *Guttulina problema*, × 150, apertural view; La Paloma

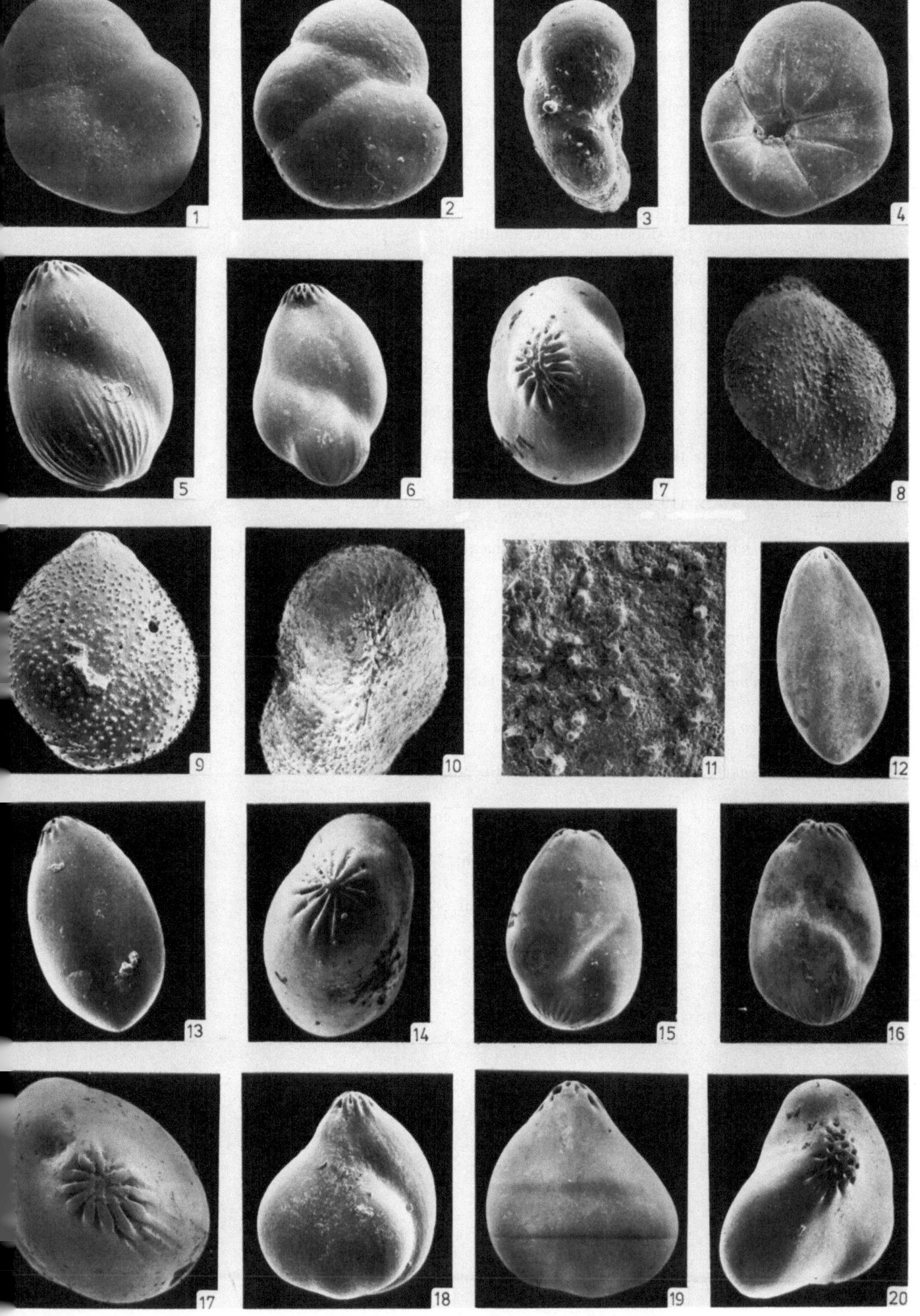

Plate 18

Fig. 1, *Gypsina vesicularis*, × 125, dorsal side; Bahía San Blas

Fig. 2, *Gypsina vesicularis*, × 125, ventral side; Bahía San Blas

Fig. 3, *Gypsina vesicularis*, × 130, edge view; Bahía San Blas

Fig. 4, *Hanzawaia boueana*, × 65, umbilical side; estuary of the Río de la Plata

Fig. 5, *Hanzawaia boueana*, × 95, umbilical side; estuary of the Río de la Plata

Fig. 6, *Hanzawaia boueana*, × 140, spiral side; estuary of the Río de la Plata

Fig. 7, *Hanzawaia boueana*, × 110, edge view; estuary of the Río de la Plata

Fig. 8, *Hanzawaia boueana*, × 200, principal and supplementary apertures; southern Brazil coast, 32°30′S

Fig. 9, *Heronallenia kempii*, × 125, spiral side; Cape Horn

Fig. 10, *Heronallenia kempii*, × 125, edge view; Cape Horn

Fig. 11, *Heronallenia kempii*, × 150, umbilical side; Cape Horn

Fig. 12, *Heronallenia kempii*, × 115, umbilical side; Tierra del Fuego

Fig. 13, *Heronallenia kempii*, × 200, aperture; Tierra del Fuego

Fig. 14, *Hoeglundina elegans*, × 85, umbilical side; Malvin Current, 34°S

Fig. 15, *Hoeglundina elegans*, × 65, spiral side; Malvin Current, 34°S

Fig. 16, *Hoeglundina elegans*, × 50, edge view; Malvin Current, 34°S

Fig. 17, *Hoeglundina elegans*, × 150, apertures; Malvin Current, 34°S

Fig. 18, *Hopkinsina pacifica*, × 250; Río de la Plata

Fig. 19, *Hopkinsina pacifica*, × 200; Río de la Plata

Fig. 20, *Hopkinsina pacifica*, × 350, apertural view; Río de la Plata

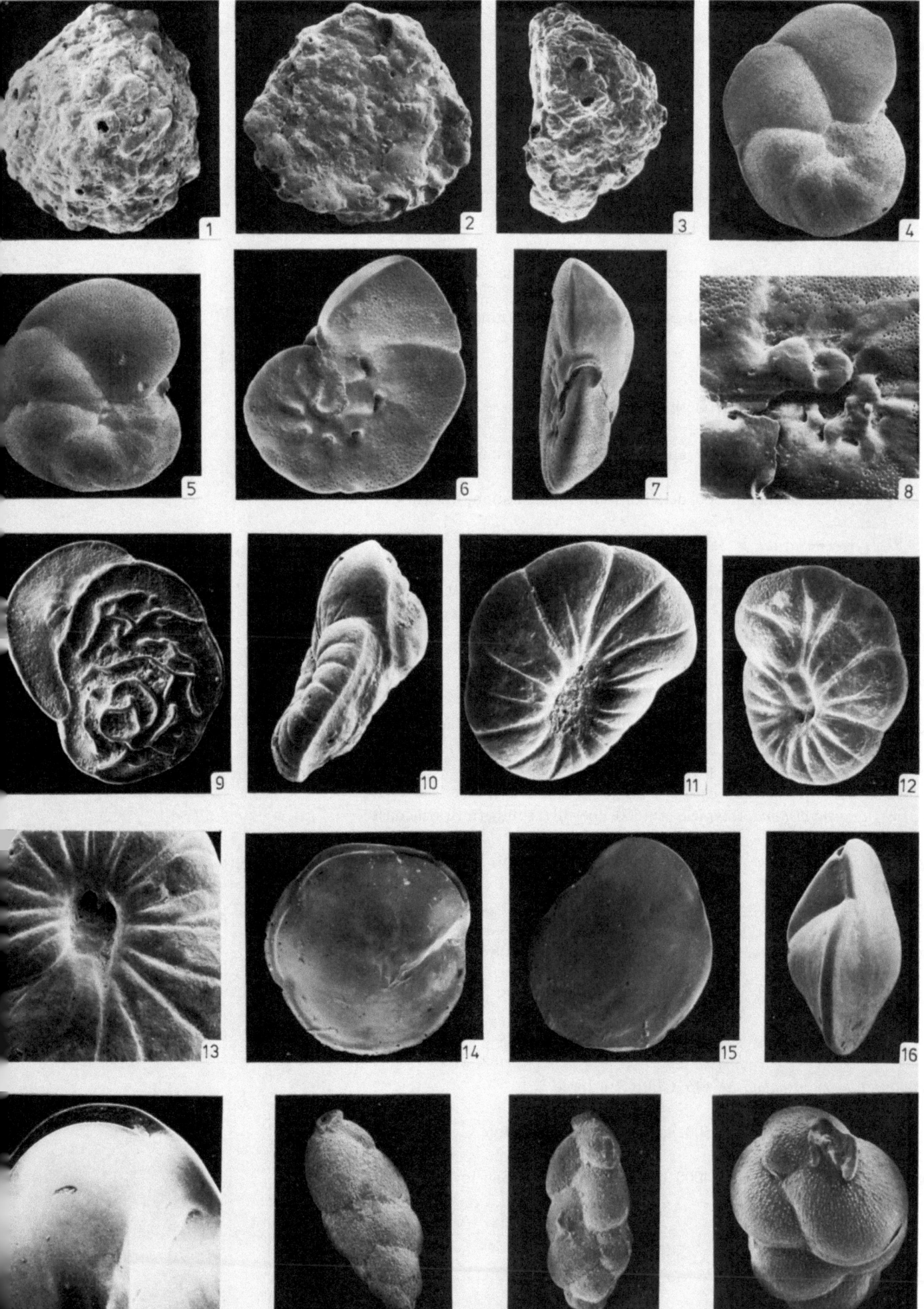

Plate 19

Fig. 1, *Lagena aspera*, × 200; Río de la Plata

Fig. 2, *Lagena aspera*, × 300, apertural view; La Paloma

Fig. 3, *Lagena aspera*, × 500, detail of wall showing pustules; Río de la Plata

Fig. 4, *Lagena caudata*, × 100, Cabo Curioso

Fig. 5, *Lagena caudata*, × 100; Malvin Current, 43°S

Fig. 6, *Lagena caudata*, × 250, aperture; Malvin Current, 36°10'S

Fig. 7, *Lagena caudata*, × 500, detail of wall ornamentation; Malvin Current, 36°10'S

Fig. 8, *Lagena clavata*, × 35; Malvin Current, 36°10'S

Fig. 9, *Lagena clavata*, × 500, aperture; Malvin Current, 36°10'S

Fig. 10, *Lagena digitale*, × 150; Puerto Deseado

Fig. 11, *Lagena digitale*, × 100; Bahía San Julián

Fig. 12, *Lagena digitale*, × 250, apertural region; Bahía San Julián

Fig. 13, *Lagena distoma*, f. typica, × 40; southern Brazil coast

Fig. 14, *Lagena distoma*, f. typica, × 1500, aperture; southern Brazil coast

Fig. 15, *Lagena distoma*, f. turgida, × 35; Malvin Current, 34°30'S

Fig. 16, *Lagena distoma*, f. turgida, × 500, aperture; Malvin Current, 34°30'S

Fig. 17, *Lagena distoma*, f. turgida, × 1500, aboral aperture; Malvin Current, 34°30'S

Fig. 18, *Lagena gracilis*, × 110; Malvin Current, 43°-44°S

Fig. 19, *Lagena gracilis*, × 125, Malvin Current, 43°-44°S

Fig. 20, *Lagena hispidula*, × 125; Golfo San Jorge

Fig. 21, *Lagena hispidula*, × 500, aperture; Golfo San Jorge

Fig. 22, *Lagena hispidula*, × 4000, detail of wall; Golfo San Jorge

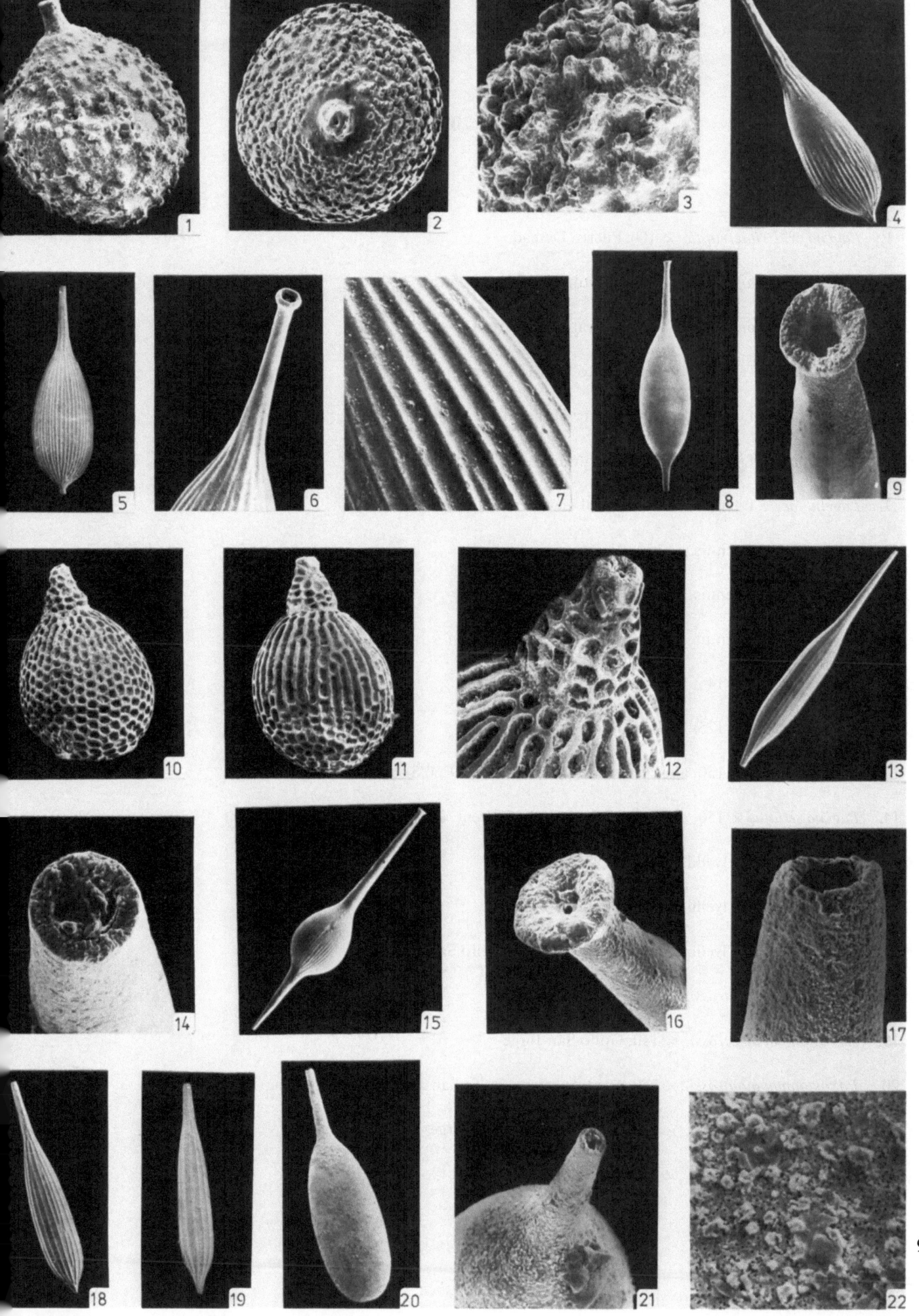

Plate 20

Fig. 1, *Lagena* cf. *L. interrupta*, × 100; Puerto Deseado

Fig. 2, *Lagena* cf. *L. interrupta*, × 250, aperture; Puerto Deseado

Fig. 3, *Lagena* cf. *L. interrupta*, × 250, aboral end; Puerto Deseado

Fig. 4, *Lagena laevis*, f. typica, × 160; Puerto Deseado

Fig. 5, *Lagena laevis*, f. typica, × 100; estuary of the Río de la Plata

Fig. 6, *Lagena laevis*, f. typica, × 250, apertural view; Río Quequén

Fig. 7, *Lagena laevis*, f. tenuis, × 250, aboral view; Río Quequén

Fig. 8, *Lagena laevis*, f. tenuis, × 150; Río Quequén

Fig. 9, *Lagena laevis*, f. tenuis, × 140; Bahía San Blas

Fig. 10, *Lagena laevis*, f. tenuis, × 1500, aperture; Río Quequén

Fig. 11, *Lagena striata*, × 145; Malvin Current, 39°30′S

Fig. 12, *Lagena striata*, × 125; Malvin Current, 36°10S

Fig. 13, *Lagena striata*, × 150, aboral view; Malvin Current, 39°30′S

Fig. 14, *Lagena striata*, × 150, apertural view; Malvin Current, 39°30′S

Fig. 15, *Lagena sulcata*, f. lyellii, × 125; Golfo San Jorge

Fig. 16, *Lagena sulcata*, f. lyellii, × 125; Golfo San Jorge

Fig. 17, *Lagena sulcata*, f. lyellii, × 200, apertural view; Golfo San Jorge

Fig. 18, *Loxostomum albatrossi*, × 200; Golfo San Jorge

Fig. 19, *Loxostomum albatrossi*, × 140; Golfo San Jorge

Fig. 20, *Loxostomum albatrossi*, × 250, apertural view; Golfo San Jorge

Fig. 21, *Loxostomum albatrossi*, × 500, aperture; Golfo San Jorge

Fig. 22, *Loxostomum albatrossi*, × 750, alveolar pits and pores; Golfo San Jorge

Plate 21

Fig. 1, *Massilina secans*, × 40; inner shelf, 34°S

Fig. 2, *Massilina secans*, × 45; inner shelf, 34°S

Fig. 3, *Massilina secans*, × 55, edge view; inner shelf, 34°S

Fig. 4, *Massilina secans*, × 150, detail of ornamentation; Río de la Plata

Fig. 5, *Melonis affine*, × 100; south of Tierra del Fuego

Fig. 6, *Melonis affine*, × 100; Malvin Current, 32°S

Fig. 7, *Melonis affine*, × 200, edge view; south of Tierra del Fuego

Fig. 8, *Miliolinella lutea*, × 100; Golfo San Jorge

Fig. 9, *Miliolinella lutea*, × 150; Golfo San Jorge

Fig. 10, *Miliolinella lutea*, × 250, aperture; Golfo San Jorge

Fig. 11, *Miliolinella subrotunda*, × 200; Ushuaia

Fig. 12, *Miliolinella subrotunda*, × 55; Ushuaia

Fig. 13, *Miliolinella subrotunda*, × 150; San Bernardo

Fig. 14, *Miliolinella subrotunda*, × 100, apertural view; Ushuaia

Fig. 15, *Morulaeplecta bulbosa*, × 110; shelf, 38°S, 57°W

Fig. 16, *Morulaeplecta bulbosa*, × 450, aperture; shelf, 38°S, 57°W

Fig. 17, *Mychostomina revertens*, × 150, spiral side; Bahía San Blas

Fig. 18, *Mychostomina revertens*, × 500, detail of spiral side; Península Valdez

Fig. 19, *Mychostomina revertens*, × 250, umbilical side; Bahía San Blas

Fig. 20, *Mychostomina revertens*, × 240, edge view; Bahía San Blas

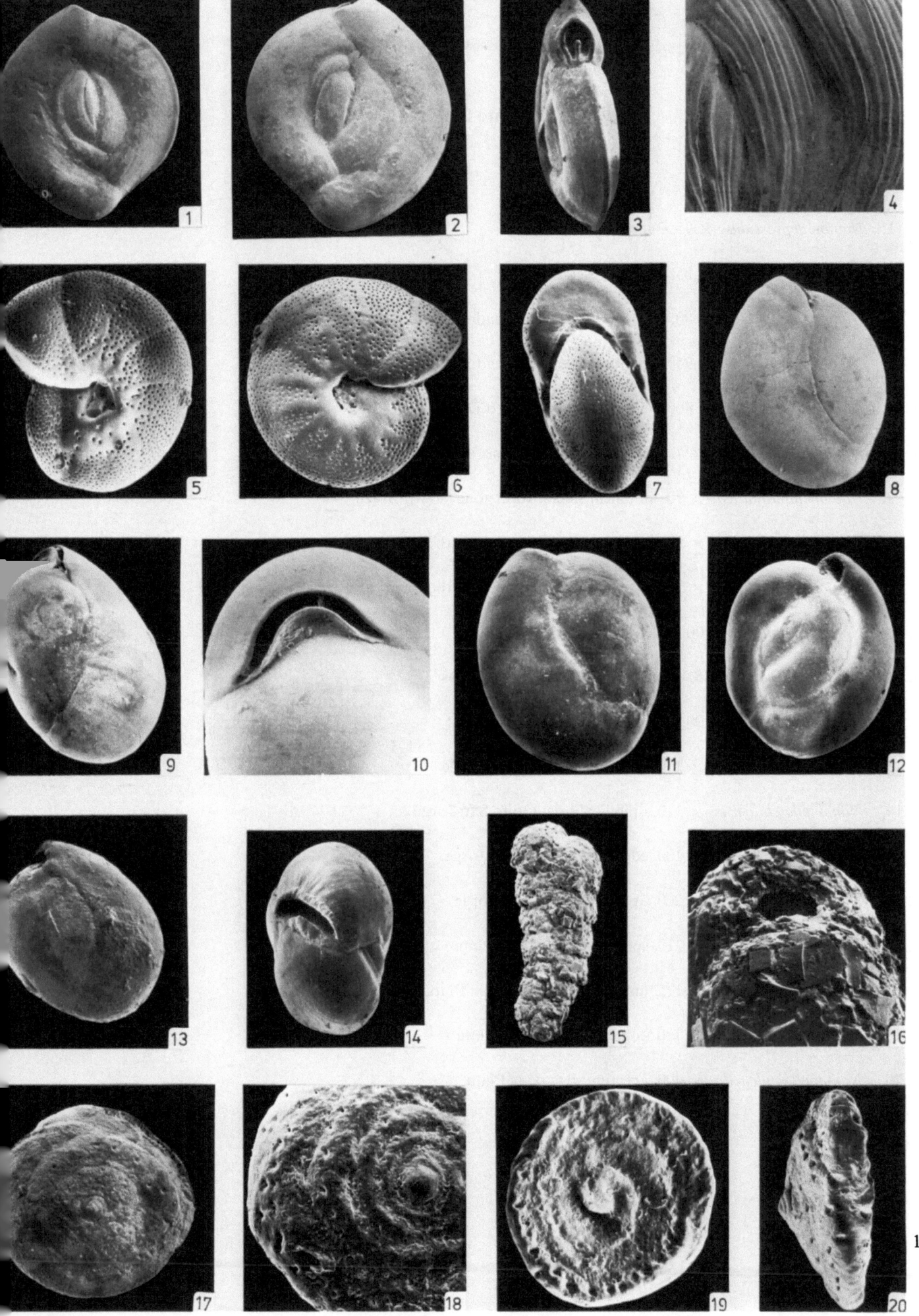

Plate 22

Fig. 1, *Nonion depressulus*, × 95; Puerto Deseado

Fig. 2, *Nonion depressulus*, × 85; Puerto Deseado

Fig. 3, *Nonion depressulus*, × 90, edge view; Puerto Deseado

Fig. 4, *Nonion depressulus*, × 250, aperture; Puerto Deseado

Fig. 5, *Nonion depressulus*, × 400, detail of suture; Puerto Deseado

Fig. 6, *Nonionella atlantica*, × 70, spiral side; littoral zone, south Brazil

Fig. 7, *Nonionella atlantica*, × 90, umbilical side; littoral zone, south Brazil

Fig. 8, *Nonionella atlantica*, × 90, edge view; littoral zone, south Brazil

Fig. 9, *Nonionella atlantica*, × 250, aperture; littoral zone, south Brazil

Fig. 10, *Nonionella auris*, × 100, spiral side; Islas Malvinas

Fig. 11, *Nonionella auris*, × 85, umbilical side; Golfo San Matías

Fig. 12, *Nonionella auris*, × 100, edge view; Golfo San Matías

Fig. 13, *Nonionella chiliensis*, × 110, spiral side; Golfo San Jorge

Fig. 14, *Nonionella chiliensis*, × 110, umbilical side; Golfo San Jorge

Fig. 15, *Nonionella chiliensis*, × 100, edge view; Golfo San Jorge

Fig. 16, *Nonionella chiliensis*, × 200, aperture; Golfo San Jorge

Fig. 17, *Nonionella pulchella*, × 210, spiral side; Río de la Plata

Fig. 18, *Nonionella pulchella*, × 175, umbilical side; Río de la Plata

Fig. 19, *Nonionella pulchella*, × 180, edge view; Río de la Plata

Fig. 20, *Nonionella pulchella*, × 400, aperture; Río de la Plata

Plate 23

Fig. 1, *Notorotalia clathrata*, × 125, spiral side; Bahía San Julián

Fig. 2, *Notorotalia clathrata*, × 145, umbilical side; Bahía San Julián

Fig. 3, *Notorotalia clathrata*, × 150, edge view; Isla de los Estados

Fig. 4, *Oolina acuticosta*, × 150; southern Brazilian coast, 32°30′S

Fig. 5, *Oolina acuticosta*, × 300, apertural view; southern Brazilian coast, 32°36′S

Fig. 6, *Oolina acuticosta*, × 200, aboral view; southern Brazilian coast, 32°30′S

Fig. 7, *Oolina borealis*, × 90; southern Brazilian coast

Fig. 8, *Oolina borealis*, × 125, apertural view; southern Brazilian coast

Fig. 9, *Oolina caudigera*, × 150; Puerto Deseado

Fig. 10, *Oolina caudigera*, × 100; Río Quequén

Fig. 11, *Oolina caudigera*, × 125, apertural view; Puerto Deseado

Fig. 12, *Oolina caudigera*, × 150, apertural view; Río Quequén

Fig. 13, *Oolina globosa*, × 160; Río de la Plata

Fig. 14, *Oolina globosa*, × 200, apertural view; Río de la Plata

Fig. 15, *Oolina hexagona*, × 200; Golfo San Jorge

Fig. 16, *Oolina hexagona*, × 200, apertural view; Golfo San Jorge

Fig. 17, *Oolina hexagona*, × 500, detail of wall ornamentation; Golfo San Jorge

Fig. 18, *Oolina lineata*, × 200; Río Gallegos

Fig. 19, *Oolina lineata*, × 250; Río Gallegos

Fig. 20, *Oolina lineata*, × 225; Río Gallegos

Fig. 21, *Oolina lineata*, × 200, apertural view; Río Gallegos

Plate 24

Fig. 1, *Oolina melo*, × 240; Puerto Deseado

Fig. 2, *Oolina melo*, × 240; Puerto Deseado

Fig. 3, *Oolina melo*, × 350, apertural view; Golfo San Jorge

Fig. 4, *Oolina melo*, × 250, aboral view; Puerto Deseado

Fig. 5, *Oolina melo*, × 500, detail of the wall; Puerto Deseado

Fig. 6, *Oolina squamosa*, × 250; Golfo San Jorge

Fig. 7, *Oolina squamosa*, × 225, apertural view; Puerto Deseado

Fig. 8, *Oolina squamosa*, × 500, detail of the wall; Golfo San Jorge

Fig. 9, *Oolina vilardeboana*, × 150, Puerto Deseado

Fig. 10, *Oolina vilardeboana*, × 125, Puerto Deseado

Fig. 11, *Oolina vilardeboana*, × 250, apertural view; Punta Medanosa

Fig. 12, *Orthomorphina calomorpha*, × 125, Puerto Deseado

Fig. 13, *Orthomorphina calomorpha*, × 400, aperture; Puerto Deseado

Fig. 14, *Orthomorphina filiformis?*, × 110, Puerto Deseado

Fig. 15, *Orthomorphina filiformis?*, × 500, aperture; Río de la Plata

Fig. 16, *Orthomorphina filiformis?*, × 500, aboral view; Río de la Plata

Fig. 17, *Patellina corrugata*, × 200, spiral view; Malvin Current, 49°08′S, 62°43′W

Fig. 18, *Patellina corrugata*, × 150, spiral side; Puerto Deseado

Fig. 19, *Patellina corrugata*, × 200, umbilical side; Golfo San Jorge

Fig. 20, *Patellina corrugata*, × 250, edge view; Ushuaia

Plate 25

Fig. 1,   *Planorbulina mediterranensis*, × 40, spiral side; southern Brazilian coast

Fig. 2,   *Planorbulina mediterranensis*, × 50, umbilical side; southern Brazilian coast

Fig. 3,   *Planorbulina mediterranensis*, × 200, apertural view; Río de la Plata

Fig. 4,   *Poroeponides lateralis*, × 40, spiral side; La Paloma

Fig. 5,   *Poroeponides lateralis*, × 60, umbilical side; La Paloma

Fig. 6,   *Poroeponides lateralis*, × 50, edge view; La Paloma

Fig. 7,   *Poroeponides lateralis*, × 150, detail of aperture and umbilicus; La Paloma

Fig. 8,   *Psammosphaera fusca*, × 50; Malvin Current, 34°34′S

Fig. 9,   *Pullenia bulloides*, × 100; continental slope east of Islas Malvinas

Fig. 10, *Pullenia bulloides*, × 100; continental slope east of Islas Malvinas

Fig. 11, *Pullenia bulloides*, × 110, apertural view; continental slope east of Islas Malvinas

Fig. 12, *Pullenia subcarinata subcarinata*, × 100; Tierra del Fuego, 54°S, 118m, *Discovery* Expd. Sta. 88, (topotypes)

Fig. 13, *Pullenia subcarinata subcarinata*, × 125, apertural view; south of Tierra del Fuego

Fig. 14, *Pullenia subcarinata quinqueloba*, × 150; inner shelf, south Brazil

Fig. 15, *Pullenia subcarinata quinqueloba*, × 145, apertural view; inner shelf, south Brazil

Fig. 16, *Pyrgo elongata*, × 100; Golfo San Jorge

Fig. 17, *Pyrgo elongata*, × 95; Bahía San Blas

Fig. 18, *Pyrgo nasuta*, × 65; Golfo San Jorge

Fig. 19, *Pyrgo nasuta*, × 65; Golfo San Jorge

Fig. 20, *Pyrgo nasuta*, × 90, apertural view; Golfo San Jorge

Fig. 21, *Pyrgo nasuta*, × 250, detail of aperture; Golfo San Jorge

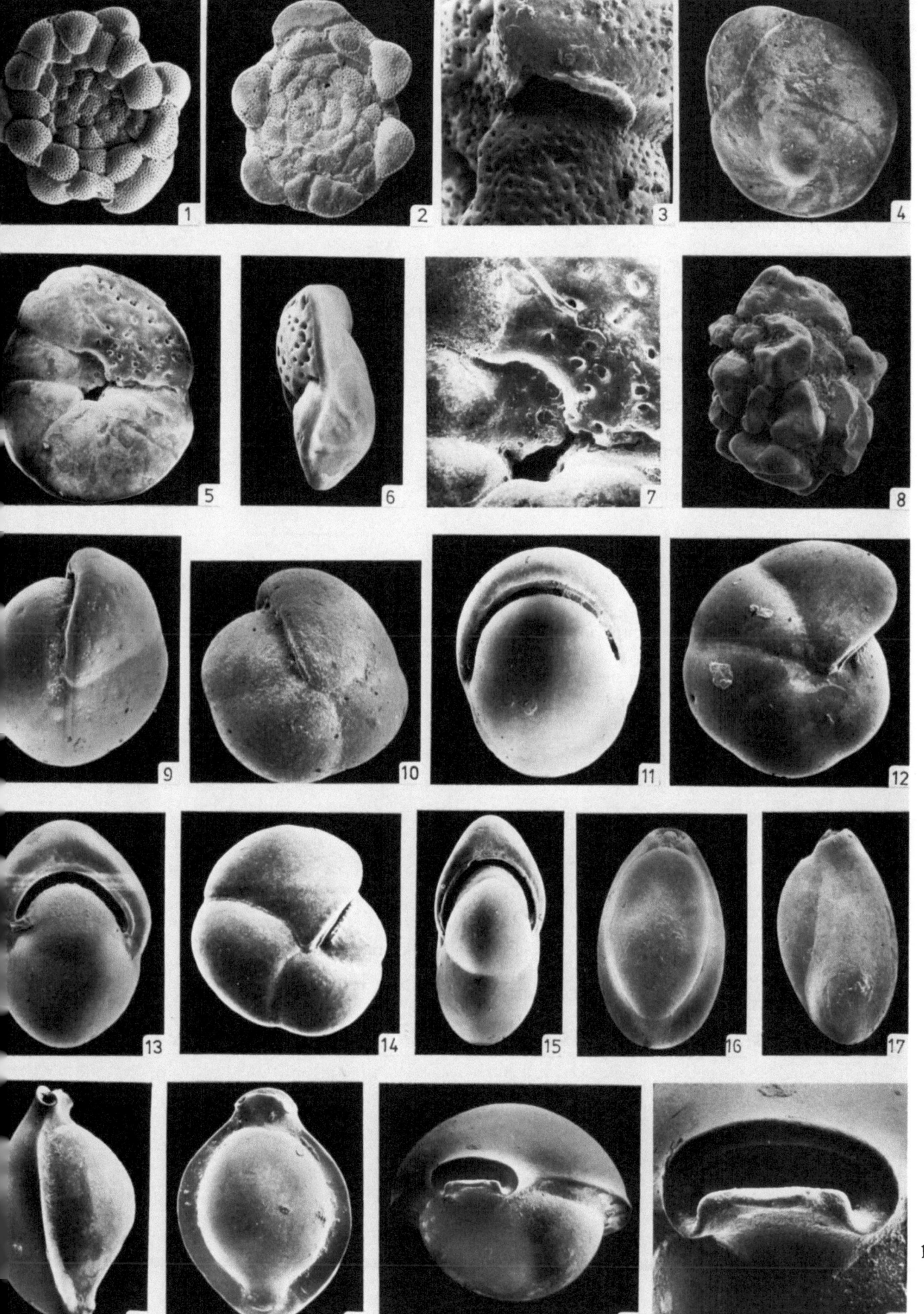

Plate 26

Fig. 1,   *Pyrgo peruviana*, × 50; south of Tierra del Fuego

Fig. 2,   *Pyrgo peruviana*, × 50; south of Tierra del Fuego

Fig. 3,   *Pyrgo peruviana*, × 50, apertural view; south of Tierra del Fuego

Fig. 4,   *Pyrgo quadrata*, × 60; Puerto Deseado

Fig. 5,   *Pyrgo quadrata*, × 50; Puerto Deseado

Fig. 6,   *Pyrgo quadrata*, × 90, apertural view; Puerto Deseado

Fig. 7,   *Pyrgo ringens*, × 100; Necochea

Fig. 8,   *Pyrgo ringens*, × 80; Necochea

Fig. 9,   *Pyrgo ringens*, × 105, apertural view; Río de la Plata

Fig. 10,   *Pyrgo subsphaerica*, × 300; Golfo San Jorge

Fig. 11,   *Pyrgo subsphaerica*, × 65; inner shelf; southern Brazil

Fig. 12,   *Pyrgo subsphaerica*, × 225; Bahía San Blas

Fig. 13,   *Pyrgo subsphaerica*, × 125, apertural view; Ostende

Fig. 14,   *Quinqueloculina angulata*, × 80; Mar del Plata

Fig. 15,   *Quinqueloculina angulata*, × 45; Mar del Plata

Fig. 16,   *Quinqueloculina angulata*, × 125, apertural view; Mar del Plata

Fig. 17,   *Quinqueloculina angulata*, × 250, detail of wall showing weak striae; San Antonio Oeste

Fig. 18,   *Quinqueloculina arctica*, × 30; Puerto Deseado

Fig. 19,   *Quinqueloculina arctica*, × 30; Puerto Deseado

Fig. 20,   *Quinqueloculina arctica*, × 50, apertural view; Puerto Deseado

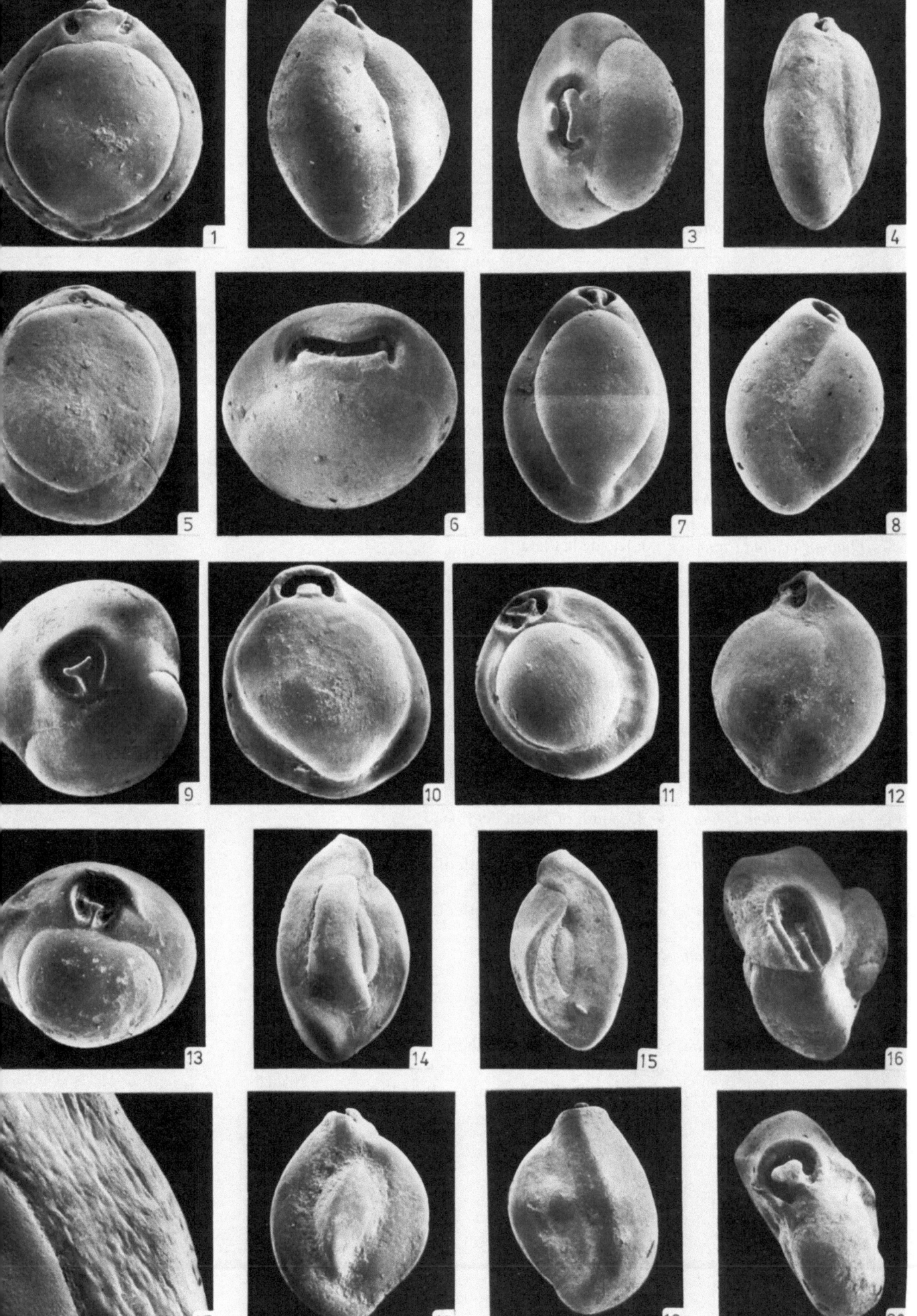

111

Plate 27

Fig. 1, *Quinqueloculina atlantica*, × 50, Uruguayan coast, 34°15′S

Fig. 2, *Quinqueloculina atlantica*, × 50; Uruguayan coast, 34°15′S

Fig. 3, *Quinqueloculina atlantica*, × 100, apertural view; Uruguayan coast, 34°15′S

Fig. 4, *Quinqueloculina brodermanni*, × 60; Uruguayan coast, 34°30′S

Fig. 5, *Quinqueloculina brodermanni*, × 55; Uruguayan coast, 34°30′S

Fig. 6, *Quinqueloculina brodermanni*, × 150, apertural view; Uruguayan coast, 34°30′S

Fig. 7, *Quinqueloculina brodermanni*, × 500, detail of wall; Uruguayan coast, 34°30′S

Fig. 8, *Quinqueloculina frigida*, × 90; Río de la Plata

Fig. 9, *Quinqueloculina frigida*, × 75; Necochea

Fig. 10, *Quinqueloculina frigida*, × 125; Río de la Plata

Fig. 11, *Quinqueloculina frigida*, × 150, apertural view; Necochea

Fig. 12, *Quinqueloculina frigida*, × 500, detail of wall; Río de la Plata

Fig. 13, *Quinqueloculina gregaria*, × 45; south of Tierra del Fuego

Fig. 14, *Quinqueloculina gregaria*, × 45; south of Tierra del Fuego

Fig. 15, *Quinqueloculina gregaria*, × 80, apertural view; south of Tierra del Fuego

Fig. 16, *Quinqueloculina gregaria*, × 90, apertural view; south of Tierra del Fuego

Fig. 17, *Quinqueloculina horrida*, × 125; Necochea

Fig. 18, *Quinqueloculina horrida*, × 110; Necochea

Fig. 19, *Quinqueloculina horrida*, × 85, apertural view; southern Brazilian coast

Fig. 20, *Quinqueloculina horrida*, × 250, detail of wall; southern Brazilian coast

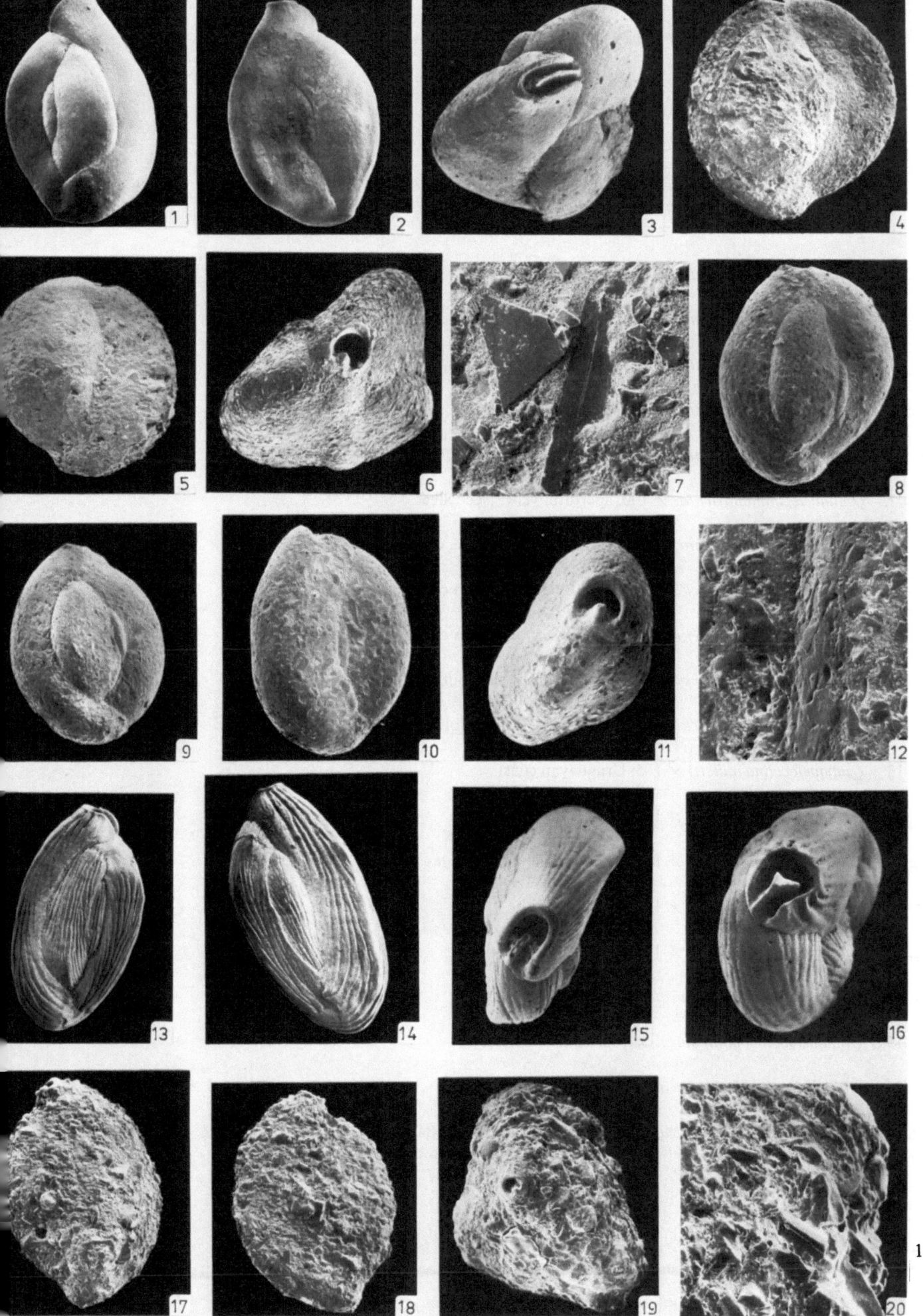

Plate 28

Fig. 1, *Quinqueloculina intricata*, × 55; Necochea

Fig. 2, *Quinqueloculina intricata*, × 80; Necochea

Fig. 3, *Quinqueloculina intricata*, × 175; Necochea

Fig. 4, *Quinqueloculina intricata*, × 50; Necochea

Fig. 5, *Quinqueloculina intricata*, × 65; Necochea

Fig. 6, *Quinqueloculina intricata*, × 125; Necochea

Fig. 7, *Quinqueloculina intricata*, × 100, apertural view; Necochea

Fig. 8, *Quinqueloculina intricata*, × 250, detail of wall showing costae; Necochea

Fig. 9, *Quinqueloculina lamarckiana*, × 75; southern Brazilian coast

Fig. 10, *Quinqueloculina lamarckiana*, × 75; southern Brazilian coast

Fig. 11, *Quinqueloculina lamarckiana*, × 80, apertural view; southern Brazilian coast

Fig. 12, *Quinqueloculina lamarckiana*, × 90, apertural view; Golfo San Jorge

Fig. 13, *Quinqueloculina milletti* × 175; Uruguayan coast

Fig. 14, *Quinqueloculina milletti* × 175; Uruguayan coast

Fig. 15, *Quinqueloculina milletti* × 150, apertural view; Uruguayan coast

Fig. 16, *Quinqueloculina milletti* × 150; Río Quequén

Fig. 17, *Quinqueloculina milletti* × 150; Río Quequén

Fig. 18, *Quinqueloculina patagonica*, × 100; Bahía San Blas

Fig. 19, *Quinqueloculina patagonica*, × 100; Bahía San Blas

Fig. 20, *Quinqueloculina patagonica*, × 100; Bahía San Blas

Fig. 21, *Quinqueloculina patagonica*, × 90, apertural view; Bahía San Blas

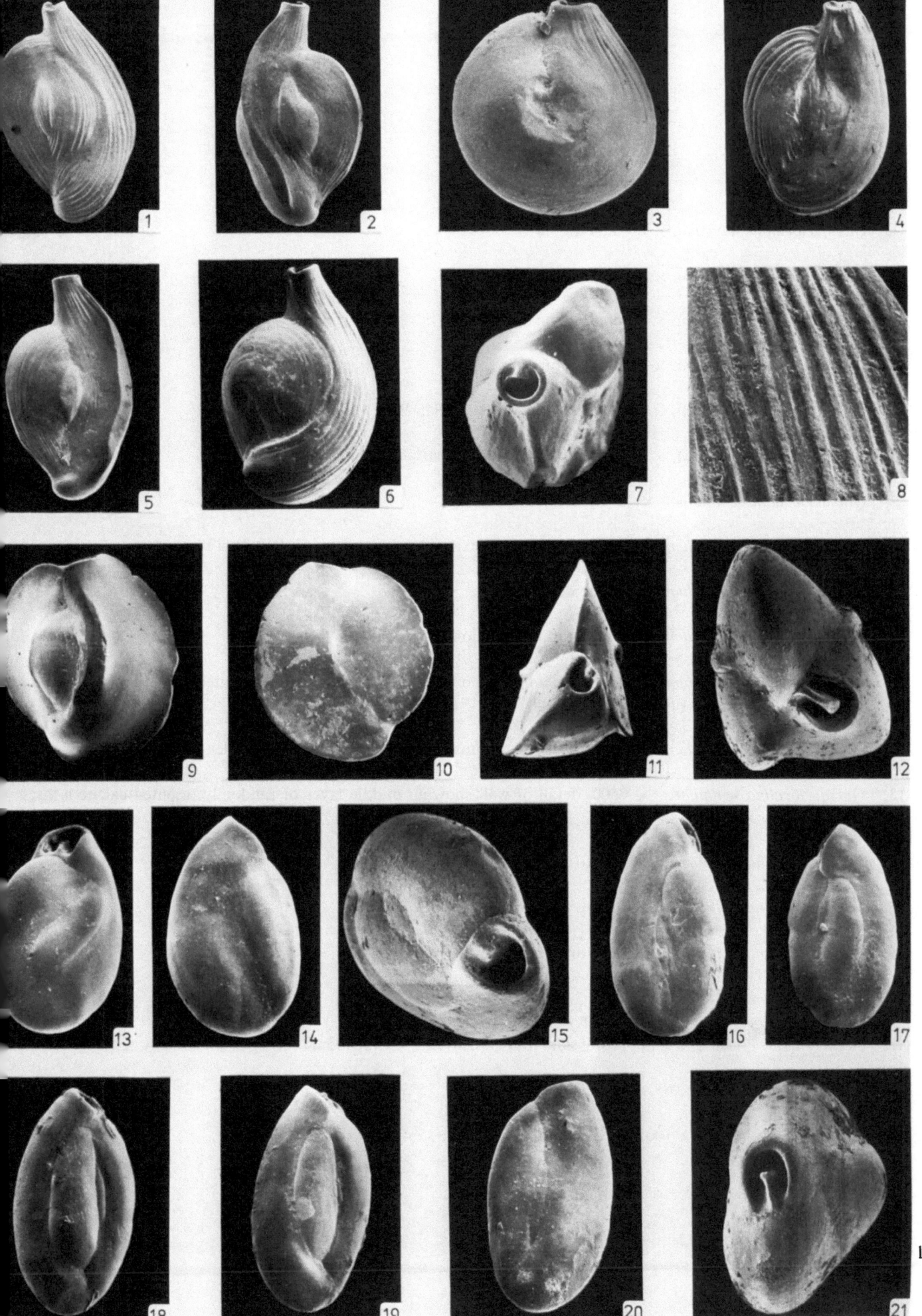

Plate 29

Fig. 1, *Quinqueloculina polygona*, × 60; La Paloma

Fig. 2, *Quinqueloculina polygona*, × 100; La Paloma

Fig. 3, *Quinqueloculina polygona*, × 75; La Paloma

Fig. 4, *Quinqueloculina polygona*, × 100, La Paloma

Fig. 5, *Quinqueloculina polygona*, × 120, apertural view; Bahía San Blas

Fig. 6, *Quinqueloculina polygona*, × 250, detail of wall showing striae and grooves; La Paloma

Fig. 7, *Quinqueloculina seminulum*, × 45; Río de la Plata

Fig. 8, *Quinqueloculina seminulum*, × 35, with protoplasm extruding from aperture; Río de la Plata

Fig. 9, *Quinqueloculina seminulum*, × 40; Río de la Plata

Fig. 10, *Quinqueloculina seminulum*, × 65, apertural view; Río de la Plata

Fig. 11, *Quinqueloculina seminulum*, × 2500, detail of wall showing relationship between oriented layer (upper left) and unoriented layer (lower right); Ushuaia

Fig. 12, *Quinqueloculina seminulum*, × 2500, detail of wall showing outer layer of oriented calcite laths; Ushuaia

Fig. 13, *Quinqueloculina seminulum*, × 2500, detail of wall showing middle layer of randomly oriented calcite laths; Ushuaia

Fig. 14, *Quinqueloculina stalkeri*, × 150; Golfo San Jorge

Fig. 15, *Quinqueloculina stalkeri*, × 150; Golfo San Jorge

Fig. 16, *Quinqueloculina stalkeri*, × 190, aperture view; Puerto Deseado

Fig. 17, *Recurvoides contortus*, × 60; Malvin Current, 35°5′S, 175m

Fig. 18, *Recurvoides contortus*, × 65, opposite side; Malvin Current, 35°5′S, 175m

Fig. 19, *Recurvoides contortus*, × 65, edge view; Malvin Current, 35°5′S

Fig. 20, *Recurvoides contortus*, × 100, aperture; Malvin Current, 35°4′S

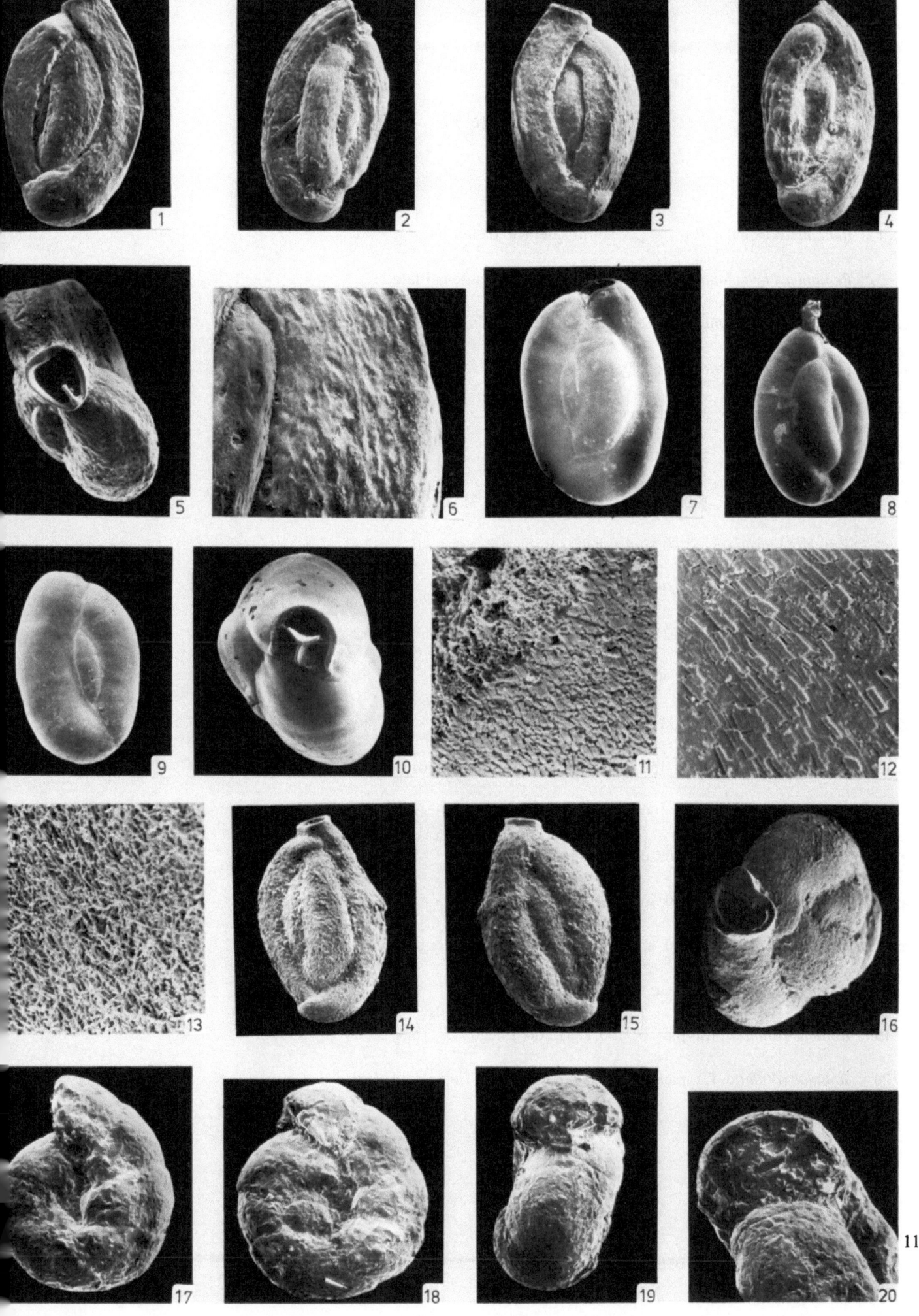

117

Plate 30

Fig. 1, *Remaneica helgolandica*, × 225, spiral side; Río de la Plata

Fig. 2, *Remaneica helgolandica*, × 175, umbilical side; Río de la Plata

Fig. 3, *Remaneica helgolandica*, × 250, umbilical side; Bahía San Julián

Fig. 4, *Remaneica helgolandica*, × 200, edge view; Puerto Deseado

Fig. 5, *Reophax curtus*, × 35; Malvin Current, 34°S

Fig. 6, *Reophax curtus*, × 30; Malvin Current, 35°30′S

Fig. 7, *Reophax curtus*, × 100, aperture; Malvin Current, 35°30′S

Fig. 8, *Reophax scorpiurus*, × 40; outer shelf, southern Brazil

Fig. 9, *Reophax scorpiurus*, × 40; outer shelf, southern Brazil

Fig. 10, *Reophax scorpiurus*, × 125, aperture; outer shelf, southern Brazil

Fig. 11, *Robulus limbosus*, s.l., × 85; southern Brazil, 32°36′S

Fig. 12, *Robulus limbosus*, s.l., × 150; Golfo San Jorge

Fig. 13, *Robulus limbosus*, s.l., × 150, apertural view; Golfo San Jorge

Fig. 14, *Robulus limbosus*, s.l., × 100, apertural view; southern Brazil

Fig. 15, *Robulus orbicularis*, × 80; south of Tierra del Fuego

Fig. 16, *Robulus orbicularis*, × 90; south of Tierra del Fuego

Fig. 17, *Robulus orbicularis*, × 90, apertural view; south of Tierra del Fuego

Fig. 18, *Robulus rotulatus*, f. typica, × 100; Golfo San Jorge

Fig. 19, *Robulus rotulatus*, f. typica, × 100; Tierra del Fuego

Fig. 20, *Robulus rotulatus*, f. typica, × 100, apertural view; Tierra del Fuego

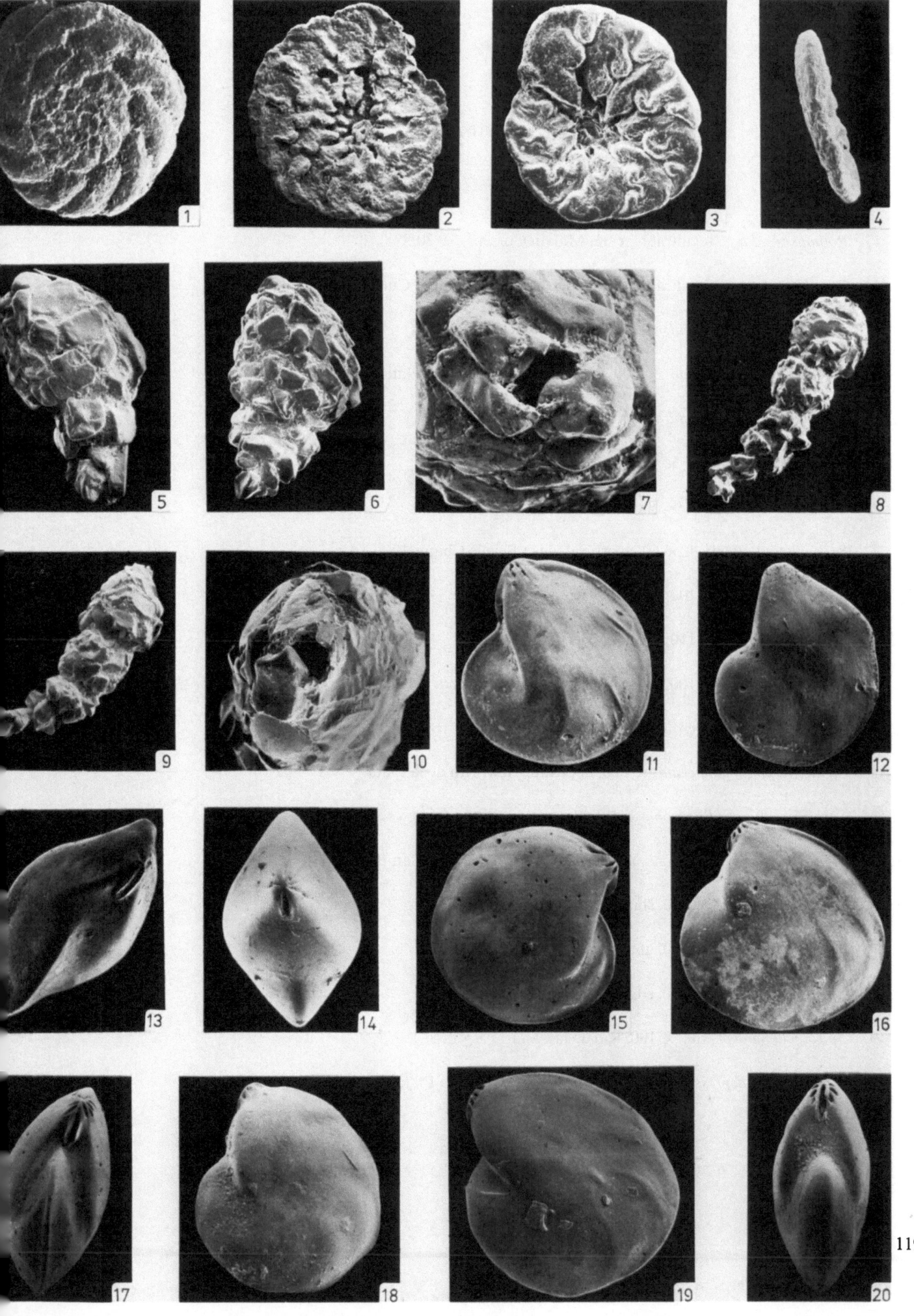

119

Plate 31

Fig. 1, *Robulus rotulatus*, f. cultrata, × 35; Malvin Current, 35°30′S

Fig. 2, *Robulus rotulatus*, f. cultrata, × 50, apertural view; Malvin Current 35°30′S

Fig. 3, *Rolshausenia rolshauseni*, × 100, spiral side; Río de la Plata

Fig. 4, *Rolshausenia rolshauseni*, × 100, umbilical side; Río de la Plata

Fig. 5, *Rolshausenia rolshauseni*, × 100, edge view; Río de la Plata

Fig. 6, *Saccammina atlantica*, × 50; Beagle Channel

Fig. 7, *Saccammina atlantica*, × 50; Beagle Channel

Fig. 8, *Saccammina atlantica*, × 60, apertural view; Beagle Channel

Fig. 9, *Sigmoilina obesa*, × 100; south of Tierra del Fuego

Fig. 10, *Sigmoilina obesa*, × 110; south of Tierra del Fuego

Fig. 11, *Sigmoilina obesa*, × 100, apertural view; south of Tierra del Fuego

Fig. 12, *Sigmomorphina pauperata*, × 200; Río de la Plata

Fig. 13, *Sigmomorphina pauperata*, × 300, apertural view; Río de la Plata

Fig. 14, *Sigmomorphina williamsoni*, × 125; Bahía San Blas

Fig. 15, *Sigmomorphina williamsoni*, × 300, apertural view; Bahía San Blas

Fig. 16, *Spirillina vivipara*, × 200, spiral side; Golfo Nuevo

Fig. 17, *Spirillina vivipara*, × 200, umbilical side; Golfo Nuevo

Fig. 18, *Spirillina vivipara*, × 300, edge view; Península Valdez

Fig. 19, *Spiroloculina depressa*, × 100; Río de la Plata

Fig. 20, *Spiroloculina depressa*, × 100, apertural view; Río de la Plata

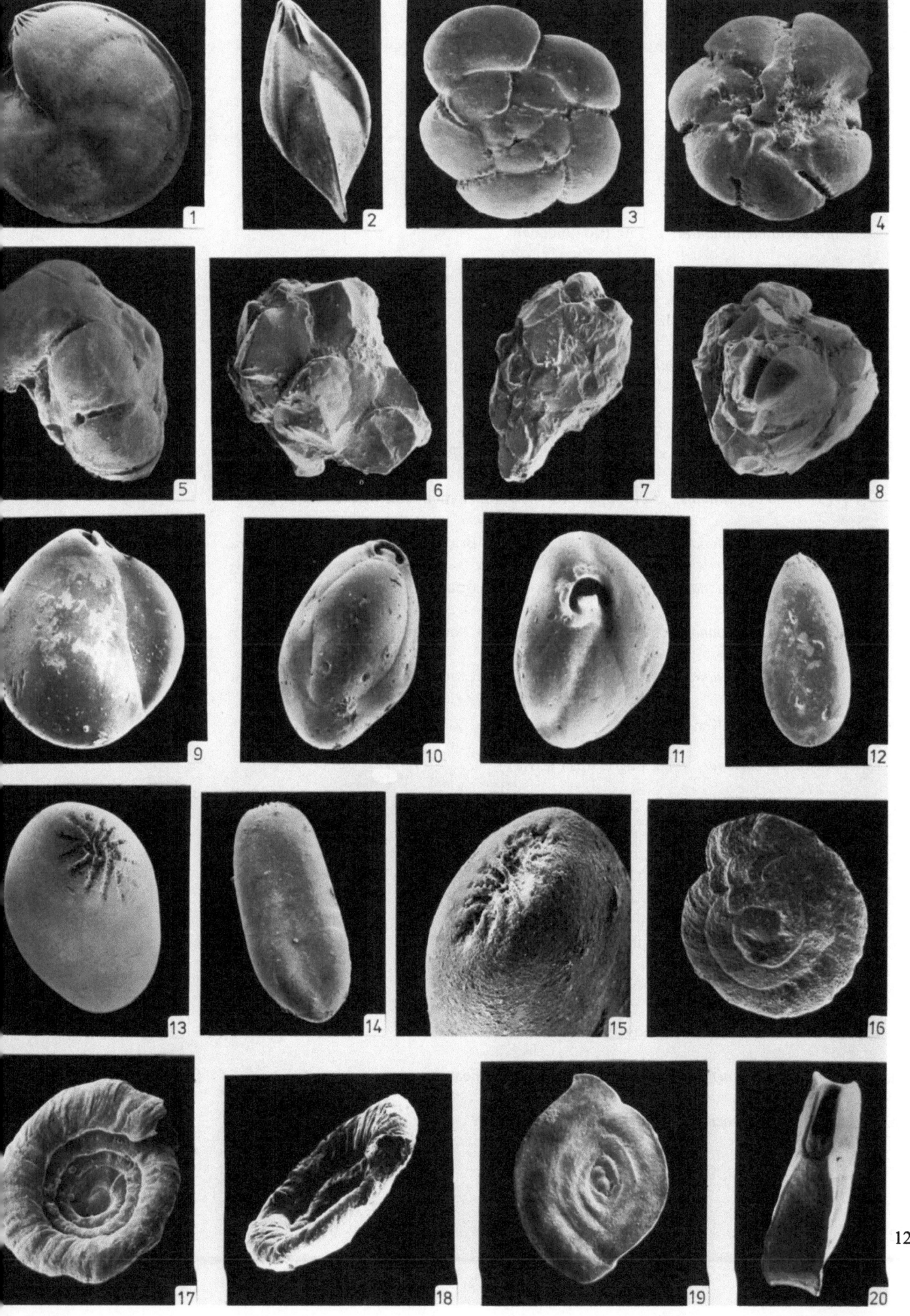

Plate 32

Fig. 1, *Spiroloculina planulata*, × 150; Río Quequén

Fig. 2, *Spiroloculina planulata*, × 150, apertural view; Río Quequén

Fig. 3, *Spiroplectammina biformis*, × 40; Ushuaia

Fig. 4, *Spiroplectammina biformis*, × 200, edge view; Malvin Current, 47°45′S, 146m

Fig. 5, *Textularia agglutinans*, × 70, edge view; inner shelf, southern Brazil

Fig. 6, *Textularia agglutinans*, × 110; inner shelf, southern Brazil

Fig. 7, *Textularia agglutinans*, × 210, apertural view; inner shelf, southern Brazil

Fig. 8, *Textularia candeiana*, × 60; inner shelf, southern Brazil

Fig. 9, *Textularia candeiana*, × 40; inner shelf, southern Brazil

Fig. 10, *Textularia candeiana*, × 50, edge view; inner shelf, southern Brazil

Fig. 11, *Textularia candeiana*, × 125, aperture; inner shelf, southern Brazil

Fig. 12, *Textularia earlandi*, × 200; Puerto Deseado

Fig. 13, *Textularia earlandi*, × 80; Lagoa dos Patos

Fig. 14, *Textularia earlandi*, × 200, edge view; Puerto Deseado

Fig. 15, *Textularia earlandi*, × 250, microspheric (?) form showing initial coil; Lagoa dos Patos

Fig. 16, *Textularia earlandi*, × 400, aperture; Lagoa dos Patos

Fig. 17, *Textularia gramen*, × 70; inner shelf, 37°11′S, 25m

Fig. 18, *Textularia gramen*, × 70; Necochea

Fig. 19, *Textularia gramen*, × 90, edge view; Necochea

Fig. 20, *Textularia gramen*, × 100, apertural view; Necochea

Fig. 21, *Textularia gramen*, × 1500, pores; Río de la Plata

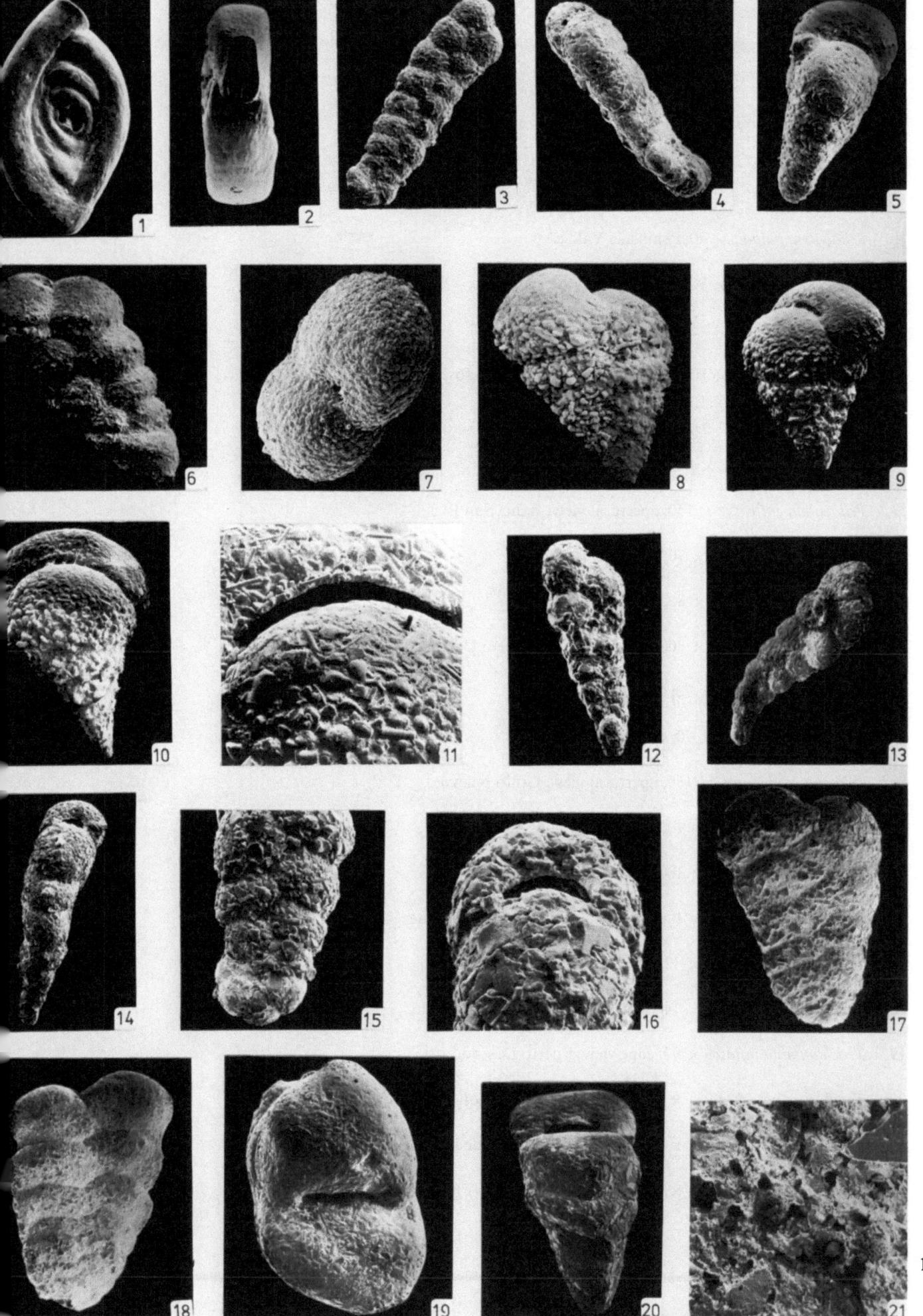

123

Plate 33

Fig. 1, *Triloculina baldai*, × 70; Península Valdez

Fig. 2, *Triloculina baldai*, × 90; Península Valdez

Fig. 3, *Triloculina baldai*, × 75; Península Valdez

Fig. 4, *Triloculina baldai*, × 100, apertural view; Golfo San José

Fig. 5, *Triloculina cultrata*, × 125; Bahía San Blas

Fig. 6, *Triloculina cultrata*, × 80; Bahía San Blas

Fig. 7, *Triloculina cultrata*, × 150, apertural view; Bahía San Blas

Fig. 8, *Triloculina laevigata*, × 70; Puerto Deseado

Fig. 9, *Triloculina laevigata*, × 85; Puerto Deseado

Fig. 10, *Triloculina laevigata*, × 200, apertural view; Puerto Deseado

Fig. 11, *Triloculina oblonga*, × 70; Golfo Nuevo

Fig. 12, *Triloculina oblonga*, × 70; Golfo Nuevo

Fig. 13, *Triloculina oblonga*, × 150, apertural view; Golfo Nuevo

Fig. 14, *Triloculina trigonula*, × 100; Península Valdez

Fig. 15, *Triloculina trigonula*, × 100; Península Valdez

Fig. 16, *Triloculina trigonula*, × 100, apertural view; Península Valdez

Fig. 17, *Trochammina inflata*, × 125, spiral side; Puerto Deseado

Fig. 18, *Trochammina inflata*, × 140, umbilical side; Puerto Deseado

Fig. 19, *Trochammina inflata*, × 90, edge view; Puerto Deseado

Fig. 20, *Trochammina ochracea*, × 225, spiral side; Río Quequén

Fig. 21, *Trochammina ochracea*, × 200, umbilical side; Río de la Plata

Fig. 22, *Trochammina ochracea*, × 220, edge view; Río de la Plata

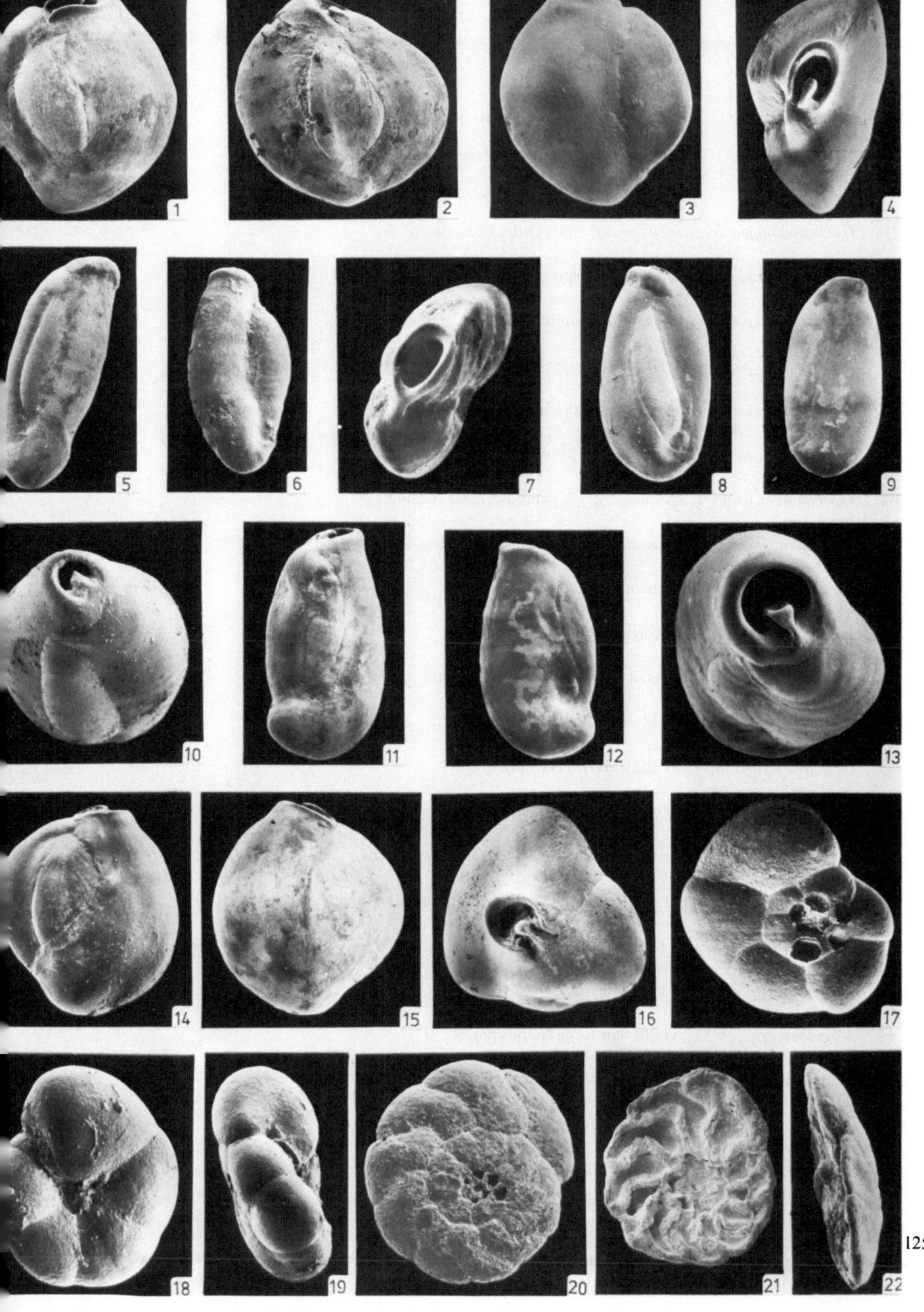

Plate 34

Fig. 1, *Trochammina plana discorbis*, × 200, spiral side; Ushuaia

Fig. 2, *Trochammina plana discorbis*, × 250, spiral side; Malvin Current, 43°S

Fig. 3, *Trochammina plana discorbis*, × 125, umbilical side; Ushuaia

Fig. 4, *Trochammina plana discorbis*, × 125, edge view; Ushuaia

Fig. 5, *Trochammina* ex gr. *T. squamata*, × 200, spiral side; Ushuaia

Fig. 6, *Trochammina* ex gr. *T. squamata*, × 175, umbilical side; Ushuaia

Fig. 7, *Trochammina* ex gr. *T. squamata*, × 175, umbilical side; Ushuaia

Fig. 8, *Trochamminu* ex gr. *T. squamata*, × 210, edge view; Puerto Deseado

Fig. 9, *Tubinella funalis*, × 40; Ushuaia

Fig. 10, *Tubinella funalis*, × 400, aperture; Puerto Deseado

Fig. 11, *Tubinella funalis*, × 350, detail of wall showing costae and suture; Ushuaia

Fig. 12, *Uvigerina bifurcata*, × 45; Malvin Current, 40°S, 98m

Fig. 13, *Uvigerina bifurcata*, × 40; Islas Malvinas (topotype)

Fig. 14, *Uvigerina bifurcata*, × 100, apertural view; Malvin Current, 40°S, 98m

Fig. 15, *Uvigerina peregrina*, f. parvula, × 100; inner shelf, southern Brazil

Fig. 16, *Uvigerina peregrina*, f. parvula, × 160; inner shelf, southern Brazil

Fig. 17, *Uvigerina striata*, × 150; Islas Malvinas

Fig. 18, *Uvigerina striata*, × 100; Islas Malvinas

Fig. 19, *Virgulina riggii*, × 135; Golfo San Jorge (paratype)

Fig. 20, *Virgulina riggii*, × 150; Golfo San Jorge (paratype)

Fig. 21, *Virgulina riggii*, × 145; Golfo San Jorge (paratype)

Fig. 22, *Virgulina riggii*, × 400, aperture; Ushuaia

127

Plate 35

Fig. 1, *Ammonia? veneta*, × 300, spiral side; Río Quequén

Fig. 2, *Ammonia? veneta*, × 300, umbilical side; Río Quequén

Fig. 3, *Ammonia? veneta*, × 350, apertural view; Río Quequén

Fig. 4, *Ammoscalaria pseudospiralis*, × 20; Uruguayan coast, 34°30′S, 83m

Fig. 5, *Ammoscalaria pseudospiralis*, × 80, apertural view; Uruguayan coast, 34°30′S, 83m

Fig. 6, *Ammoscalaria pseudospiralis*, × 250, aperture of the intercameral septa; Río de la Plata

Fig. 7, *Ammoscalaria pseudospiralis*, × 40, marginal view; Uruguayan coast, 34°51′S

Fig. 8, *Ammoscalaria tenuimargo*, × 50; Río de la Plata

Fig. 9, *Ammoscalaria tenuimargo*, × 60; southern Brazil shelf

Fig. 10, *Ammoscalaria tenuimargo*, × 70, marginal view; southern Brazil shelf

Fig. 11, *Ammoscalaria tenuimargo*, × 80, apertural view; Río de la Plata

Fig. 12, *Ammotium cassis*, × 225; Lagoa dos Patos

Fig. 13, *Ammotium cassis*, × 500, apertural view; Lagoa dos Patos

Fig. 14, *Ammotium salsum*, × 125; Lagoa dos Patos

Fig. 15, *Ammotium salsum*, × 150; Lagoa dos Patos

Fig. 16, *Ammotium salsum*, × 175; Uruguayan coast of the Río de la Plata

Fig. 17, *Ammotium salsum*, × 160, apertural view; Lagoa dos Patos

Fig. 18, *Arenoparrella mexicana*, × 125, spiral side; Lagoa dos Patos

Fig. 19, *Arenoparrella mexicana*, × 110, umbilical side; Lagoa dos Patos

Fig. 20, *Arenoparrella mexicana*, × 125, apertural view; Lagoa dos Patos

Fig. 21, *Arenoparrella mexicana*, × 125, apertural view; Lagoa dos Patos

129

Plate 36

Fig. 1, *Haplophragmoides wilberti*, × 150; Lagoa dos Patos

Fig. 2, *Haplophragmoides wilberti*, × 175, edge view; Lagoa dos Patos

Fig. 3, *Haplophragmoides wilberti*, × 150, edge view; Lagoa dos Patos

Fig. 4, *Haplophragmoides wilberti*, × 400, aperture; Lagoa dos Patos

Fig. 5, *Jadammina polystoma*, × 100, spiral side; Puerto Deseado

Fig. 6, *Jadammina polystoma*, × 110, umbilical side; Río Quequén

Fig. 7, *Jadammina polystoma*, × 90, edge view; Puerto Deseado

Fig. 8, *Miliammina fusca*, × 90; Ushuaia

Fig. 9, *Miliammina fusca*, × 90; Ushuaia

Fig. 10, *Miliammina fusca*, × 100; Golfo San Jorge

Fig. 11, *Miliammina fusca*, × 200, apertural view; Puerto Deseado

Fig. 12, *Miliammina fusca*, × 225, apertural view; Bahía San Julían

Fig. 13, *Nonion? pseudotisburyense*, × 250; delta of the Río Paraná

Fig. 14, *Nonion? pseudotisburyense*, × 250; delta of the Río Paraná

Fig. 15, *Nonion? pseudotisburyense*, × 175; delta of the Río Paraná

Fig. 16, *Nonion? pseudotisburyense*, × 300, juvenile; delta of the Río Paraná

Fig. 17, *Nonion? pseudotisburyense*, × 225, edge view; delta of the Río Paraná

Fig. 18, *Nonion? pseudotisburyense*, × 225, edge view; delta of the Río Paraná

Fig. 19, *Nonion? pseudotisburyense*, × 200, edge view; delta of the Río Paraná

Fig. 20, *Protoschista findens*, × 90; Lagoa dos Patos

Fig. 21, *Protoschista findens*, × 425, apertural view; Lagoa dos Patos

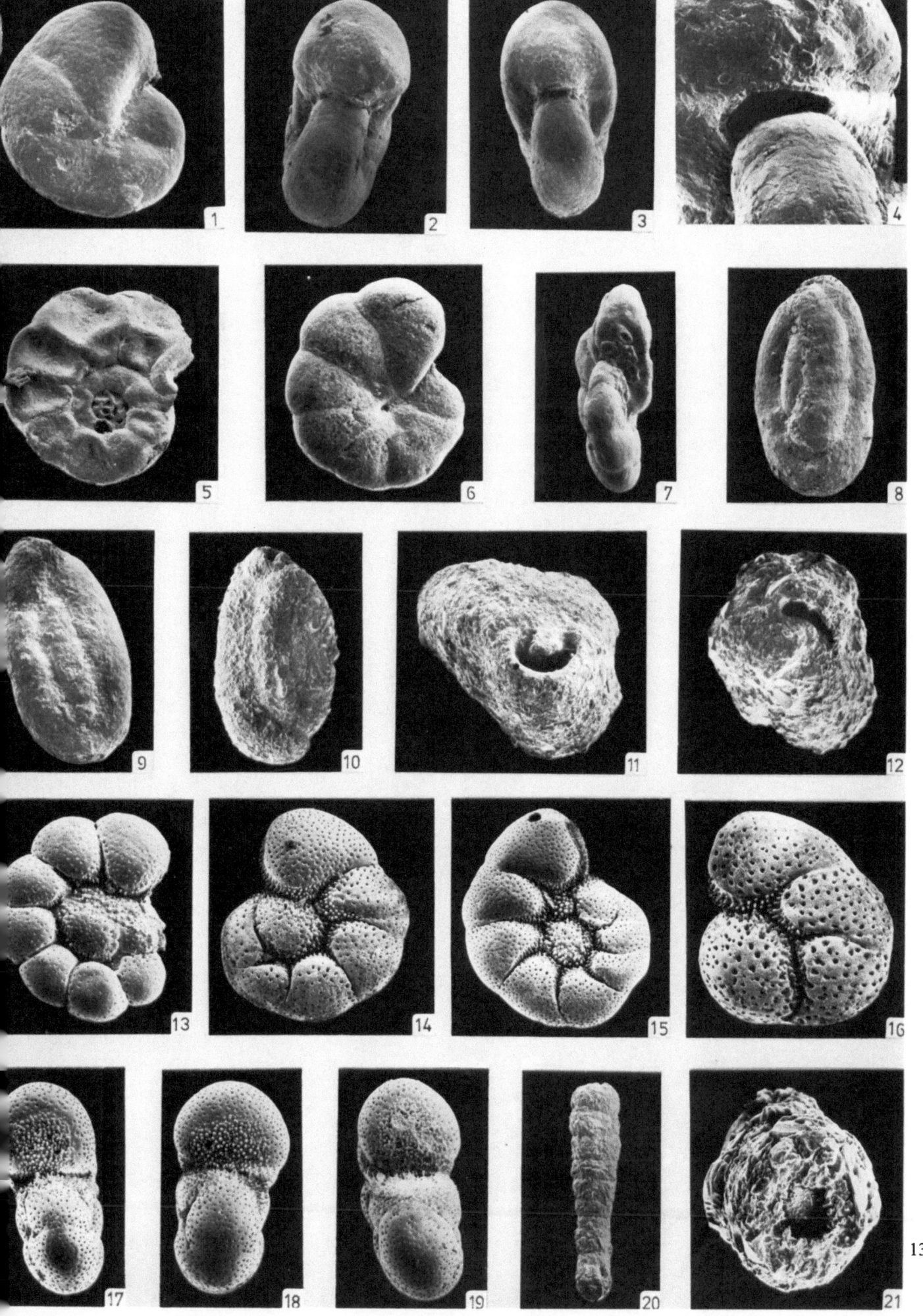

MAPS

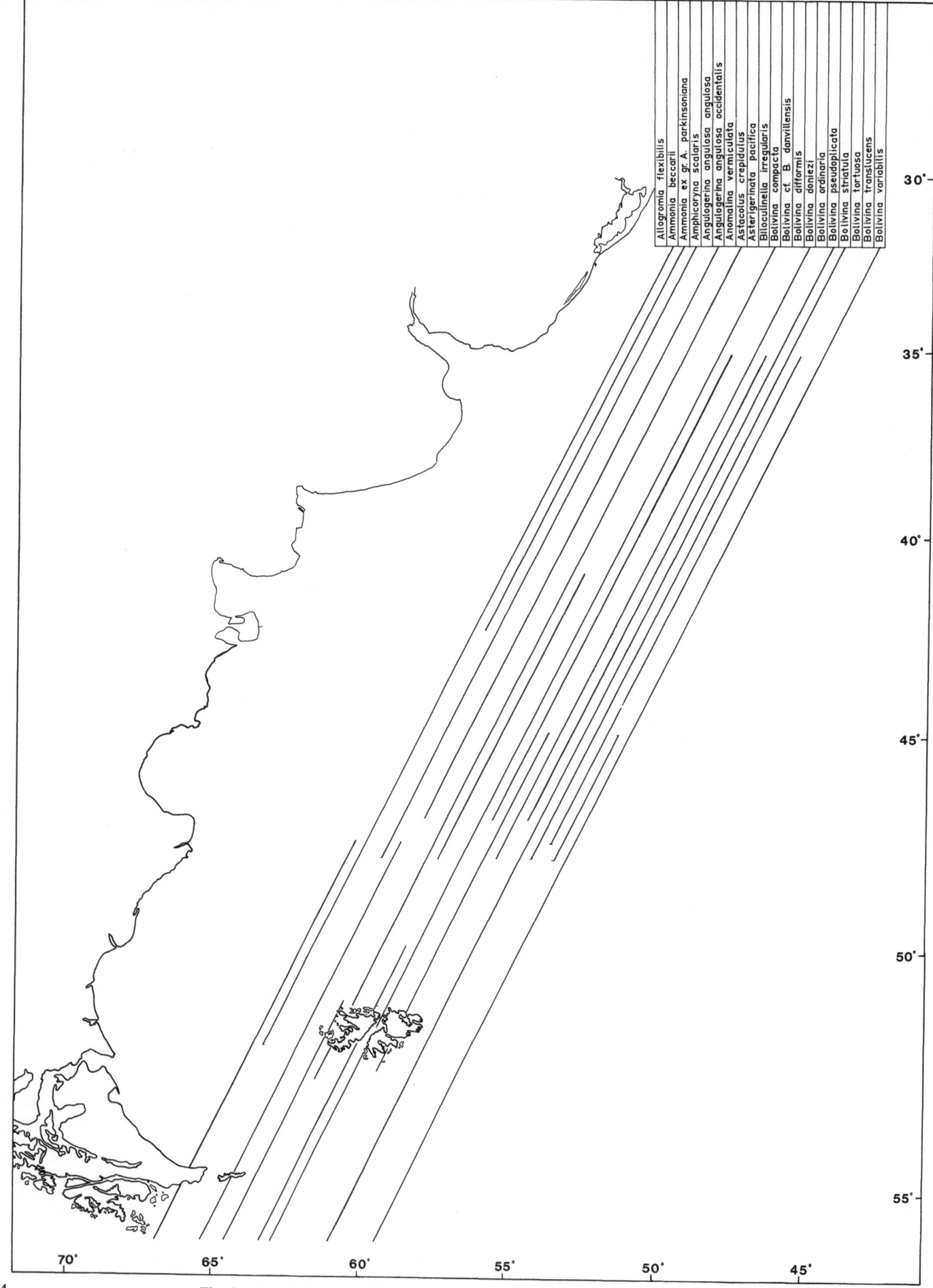

The following species labels appear in the legend box:

Allogromia flexibilis
Ammonia beccarii
Ammonia ex gr. A. parkinsoniana
Amphicoryna scalaris
Angulogerina angulosa angulosa
Angulogerina angulosa occidentalis
Anomalina vermiculata
Astacolus crepidulus
Asterigerinata pacifica
Biloculinella irregularis
Bolivina compacta
Bolivina cf. B. danvillensis
Bolivina difformis
Bolivina doniezi
Bolivina ordinaria
Bolivina pseudoplicata
Bolivina striatula
Bolivina tortuosa
Bolivina translucens
Bolivina variabilis

134   Fig. 4.   Distribution of shelf species (*Allogromia flexibilis – Bolivina variabilis*)

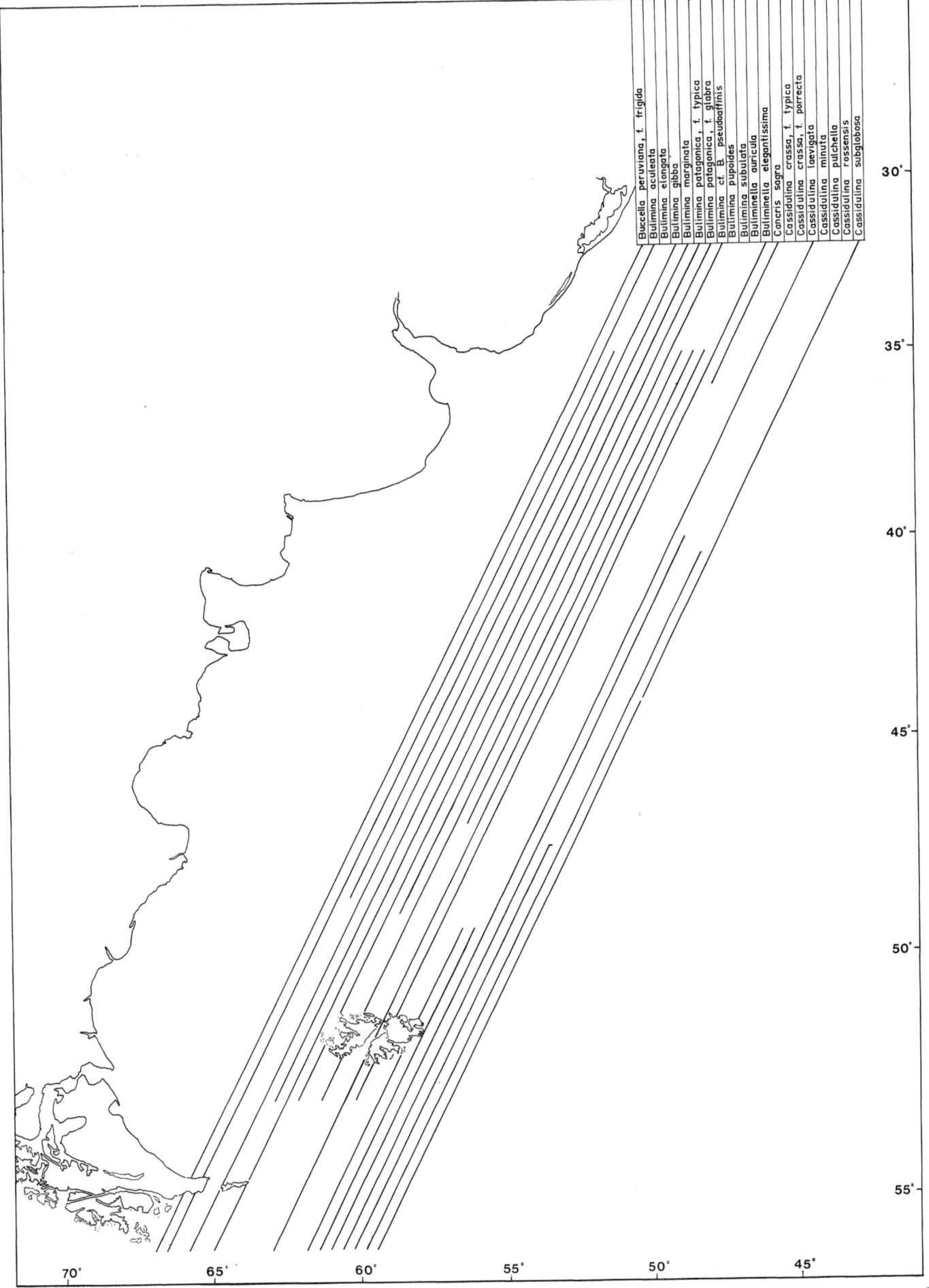

Fig. 5. Distribution of shelf species (*Buccella peruviana* – *Cassidulina subglobosa*)

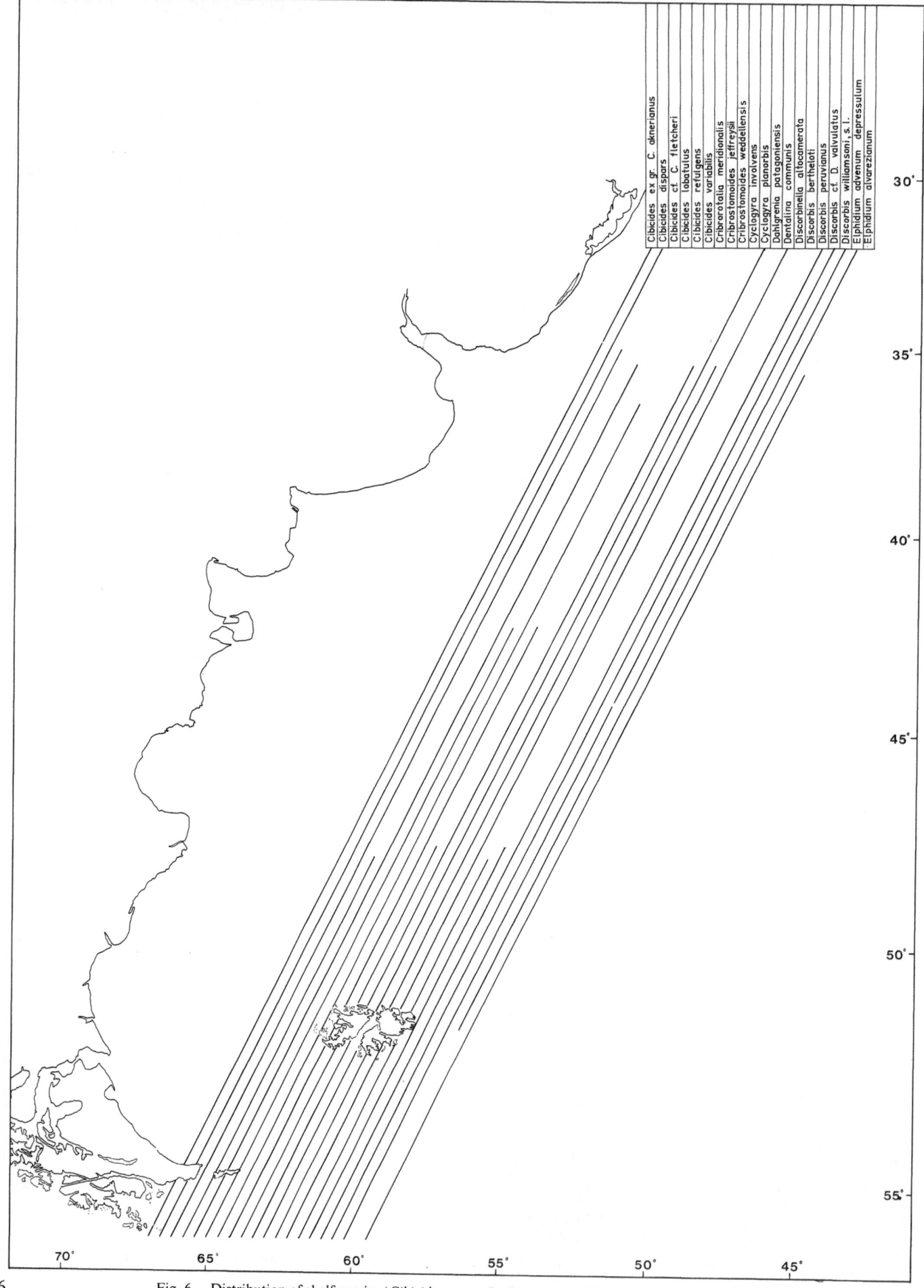

Fig. 6. Distribution of shelf species (*Cibicides* ex gr. *C. aknerianus* – *Elphidium alvarezianum*)

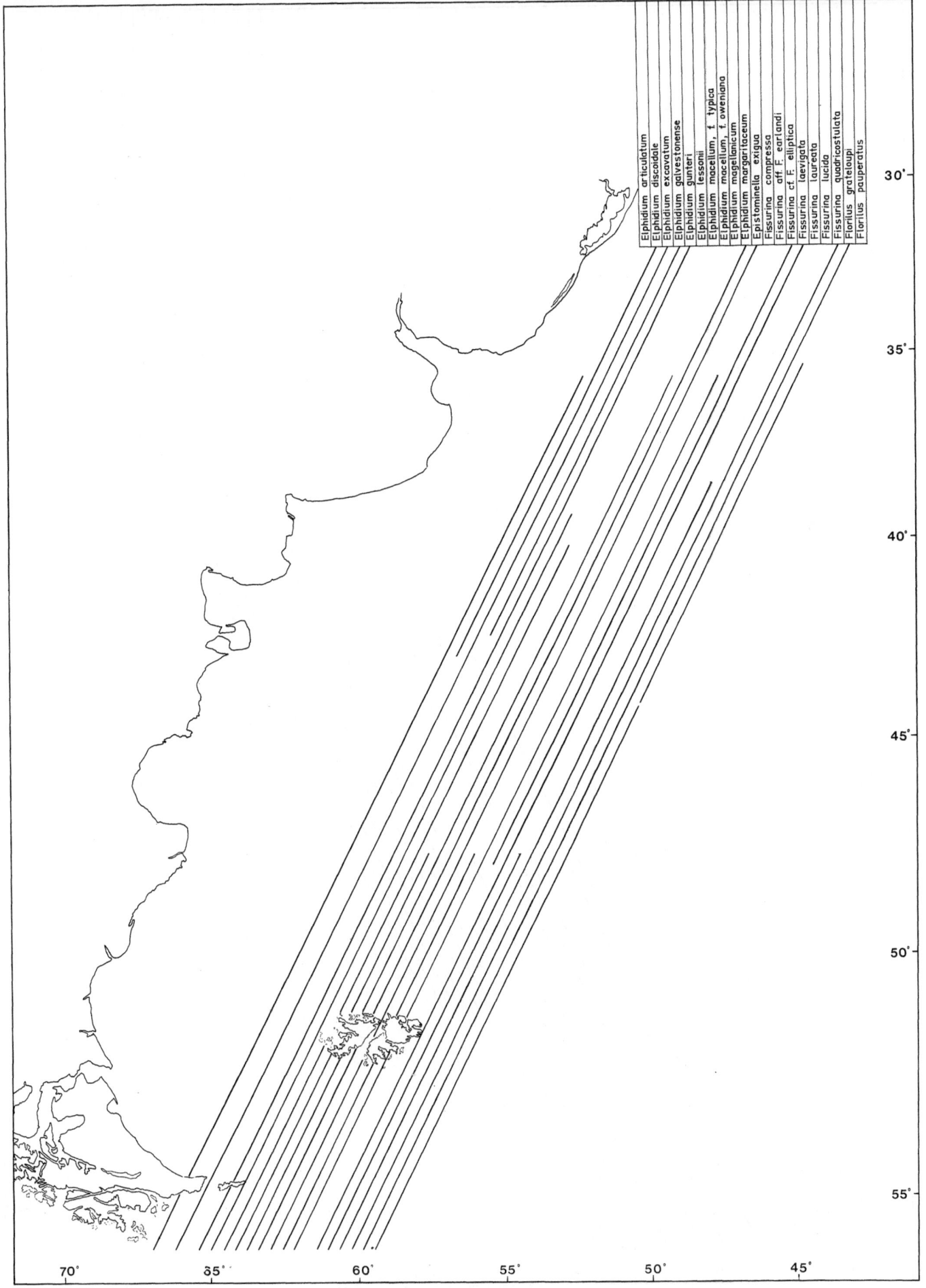

Fig. 7. Distribution of shelf species (*Elphidium articulatum – Florilus pauperatus*)

137

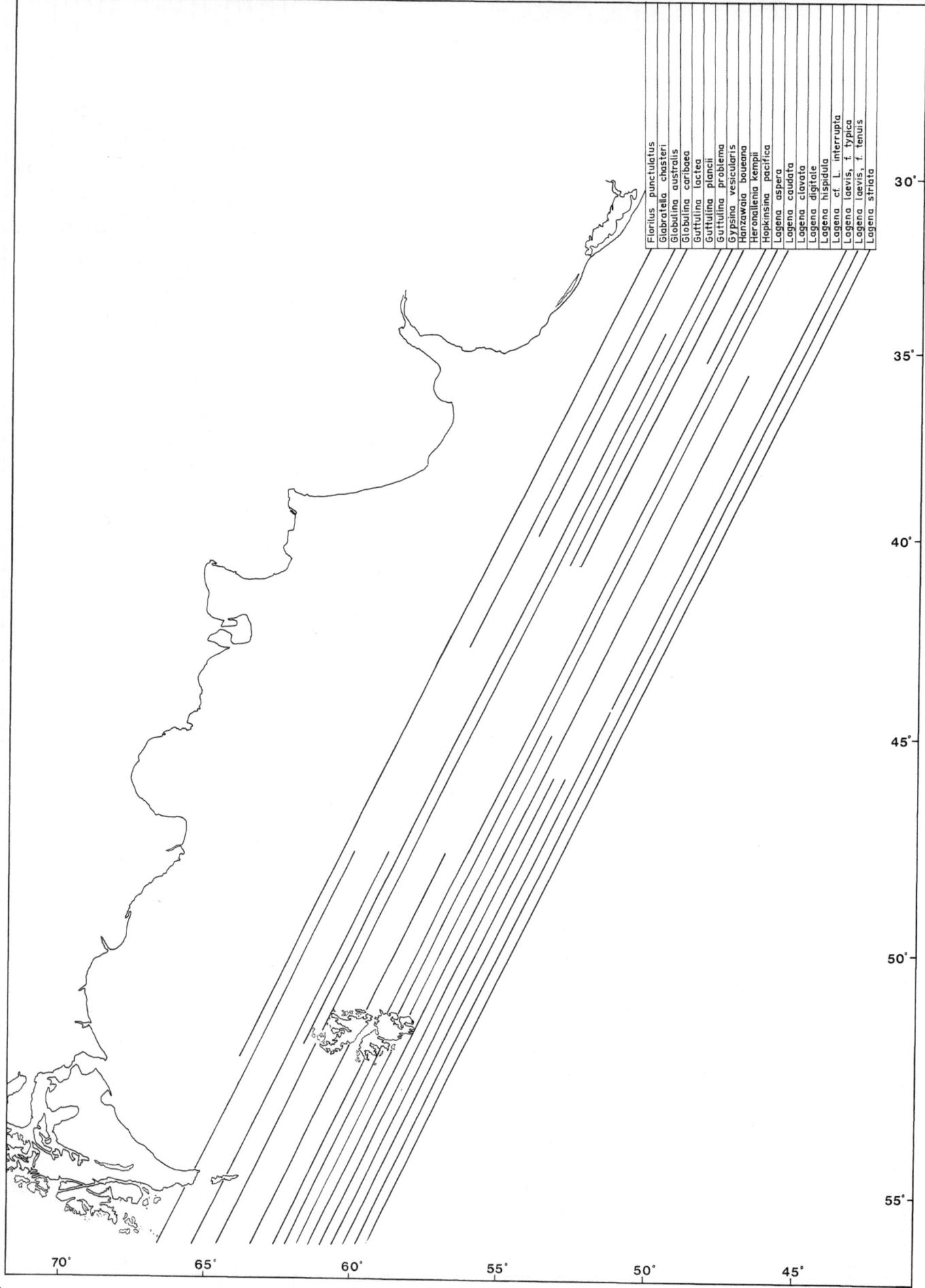

Florilus punctulatus
Glabratella chasteri
Globulina australis
Globulina caribaea
Guttulina lactea
Guttulina plancii
Guttulina problema
Gypsina vesicularis
Hanzawaia boueana
Heronallenia kempii
Hopkinsina pacifica
Lagena aspera
Lagena caudata
Lagena clavata
Lagena digitale
Lagena hispidula
Lagena cf. L. interrupta
Lagena laevis, f. typica
Lagena laevis, f. tenuis
Lagena striata

Fig. 8.   Distribution of shelf species (*Florilus punctulatus – Lagena striata*)

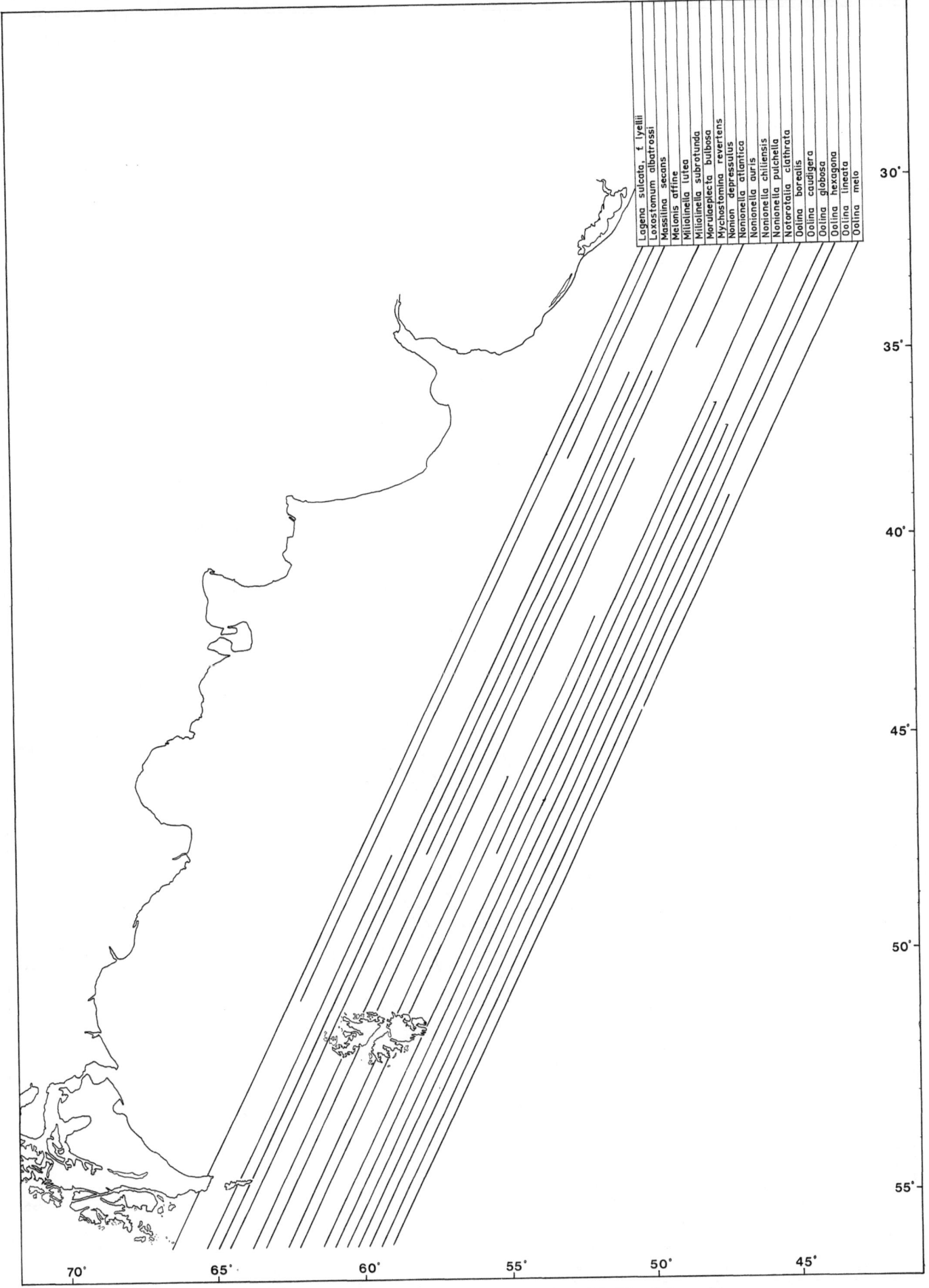

Fig. 9. Distribution of shelf species (*Lagena sulcata*, f. lyellii – *Oolina melo*)

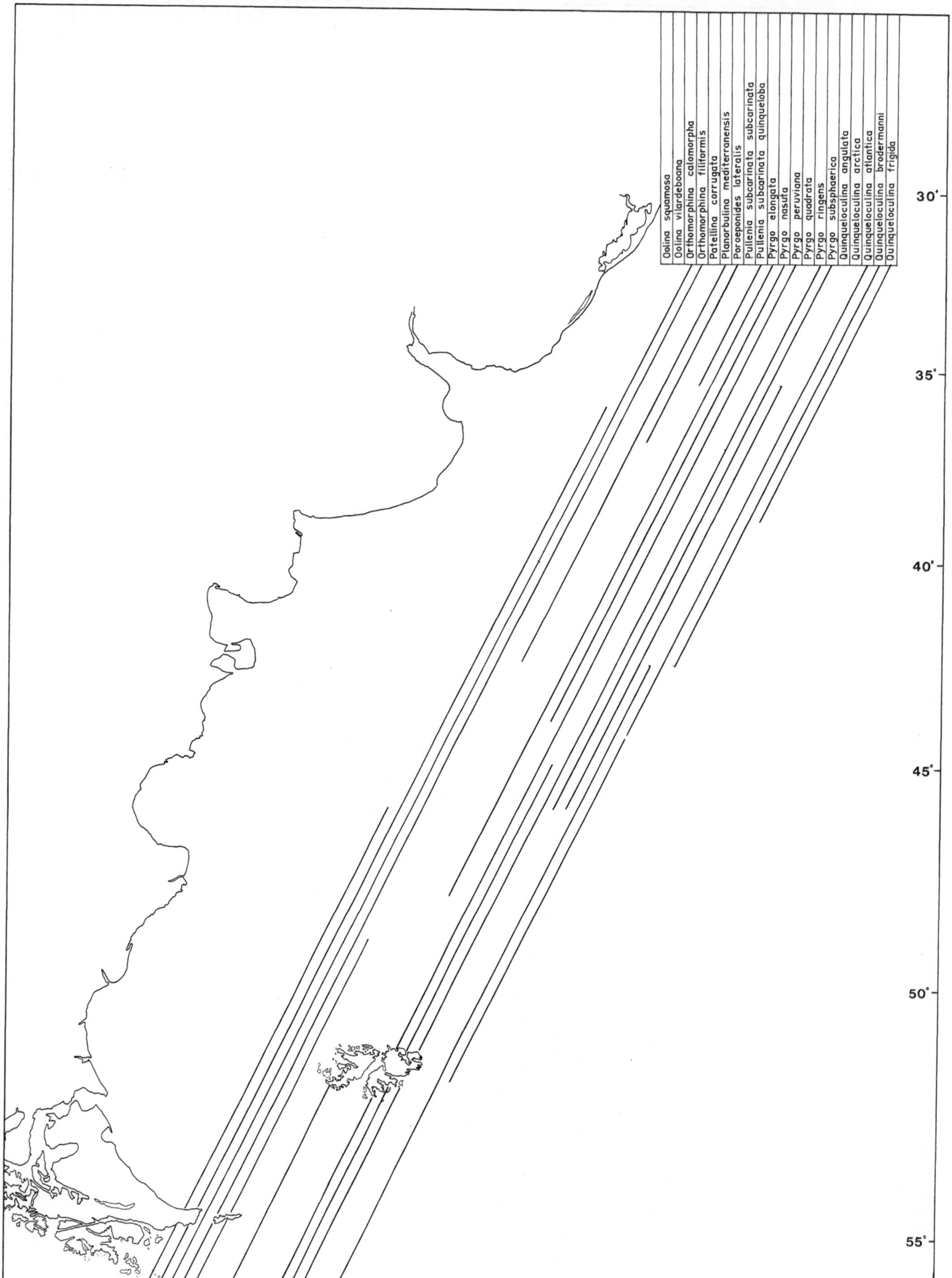
Oolina squamosa
Oolina vilardeboana
Orthomorphina calomorpha
Orthomorphina filiformis
Patellina corrugata
Planorbulina mediterranensis
Poroeponides lateralis
Pullenia subcarinata subcarinata
Pullenia subcarinata quinqueloba
Pyrgo elongata
Pyrgo nasuta
Pyrgo peruviana
Pyrgo quadrata
Pyrgo ringens
Pyrgo subsphaerica
Quinqueloculina angulata
Quinqueloculina arctica
Quinqueloculina atlantica
Quinqueloculina brodermanni
Quinqueloculina frigida
30°
35°
40°
45°
50°
55°
70°
65°
60°
55°
50°
45°

Fig. 10. Distribution of shelf species (*Oolina squamosa – Quinqueloculina frigida*)

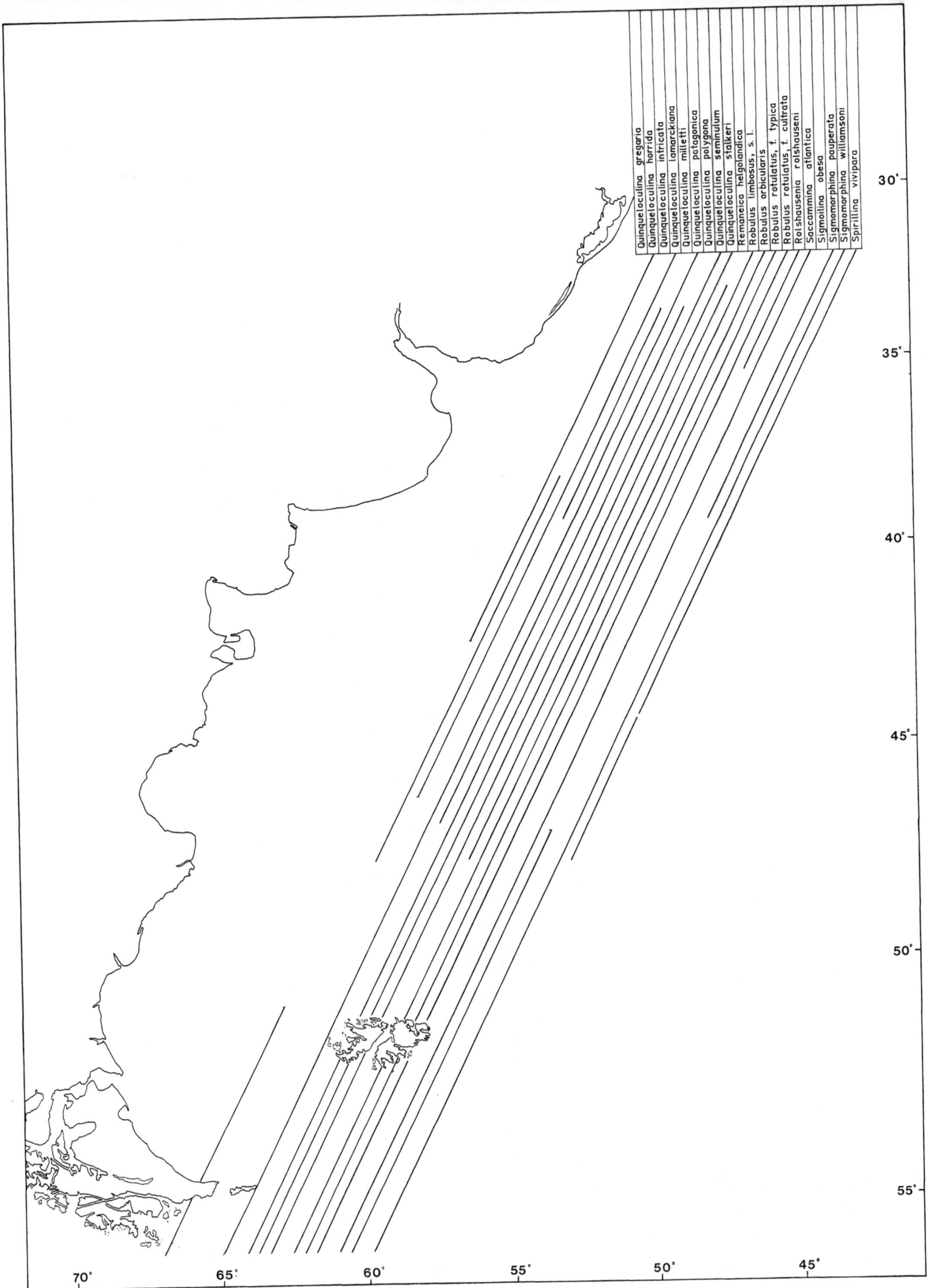
Quinqueloculina gregaria
Quinqueloculina horrida
Quinqueloculina intricata
Quinqueloculina lamarckiana
Quinqueloculina milletti
Quinqueloculina patagonica
Quinqueloculina polygona
Quinqueloculina seminulum
Quinqueloculina stalkeri
Remaneica helgolandica
Robulus limbosus, s. l.
Robulus orbicularis
Robulus rotulatus, f. typica
Robulus rotulatus, f. cultrata
Rolshausenia rolshauseni
Saccammina atlantica
Sigmoilina obesa
Sigmomorphina pauperata
Sigmomorphina williamsoni
Spirillina vivipara
30°
35°
40°
45°
50°
55°
70°
65°
60°
55°
50°
45°

Fig. 11. Distribution of shelf species (*Quinqueloculina gregaria – Spirillina vivipara*)

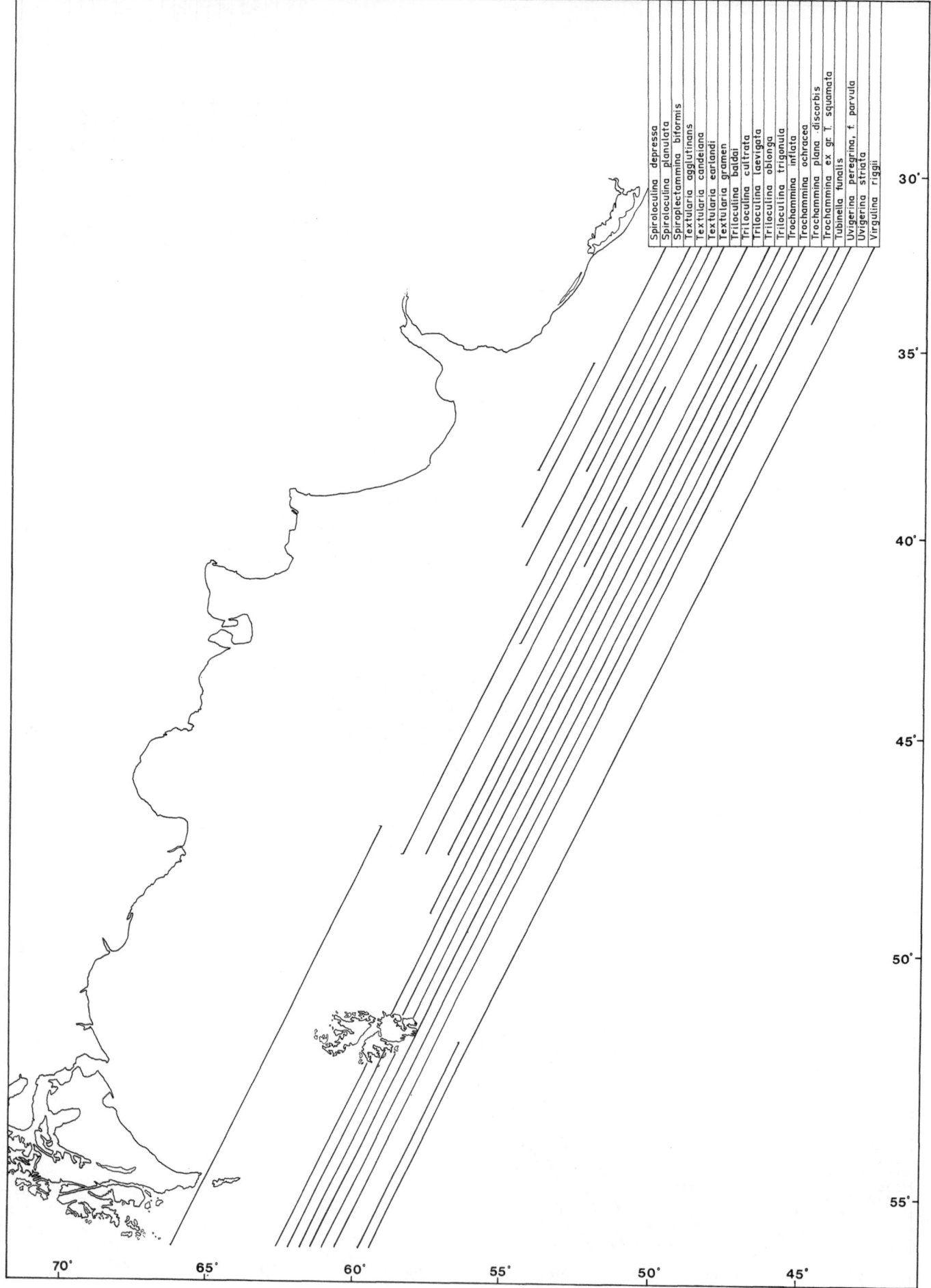

Spiroloculina depressa
Spiroloculina planulata
Spiroplectammina biformis
Textularia agglutinans
Textularia candeiana
Textularia earlandi
Textularia gramen
Triloculina baidai
Triloculina cultrata
Triloculina laevigata
Triloculina oblonga
Triloculina trigonula
Trochammina inflata
Trochammina ochracea
Trochammina plana discorbis
Trochammina ex gr. T. squamata
Tubinella funalis
Uvigerina peregrina, f. parvula
Uvigerina striata
Virgulina riggii

Fig. 12.   Distribution of shelf species (*Spiroloculina depressa* – *Virgulina riggii*)

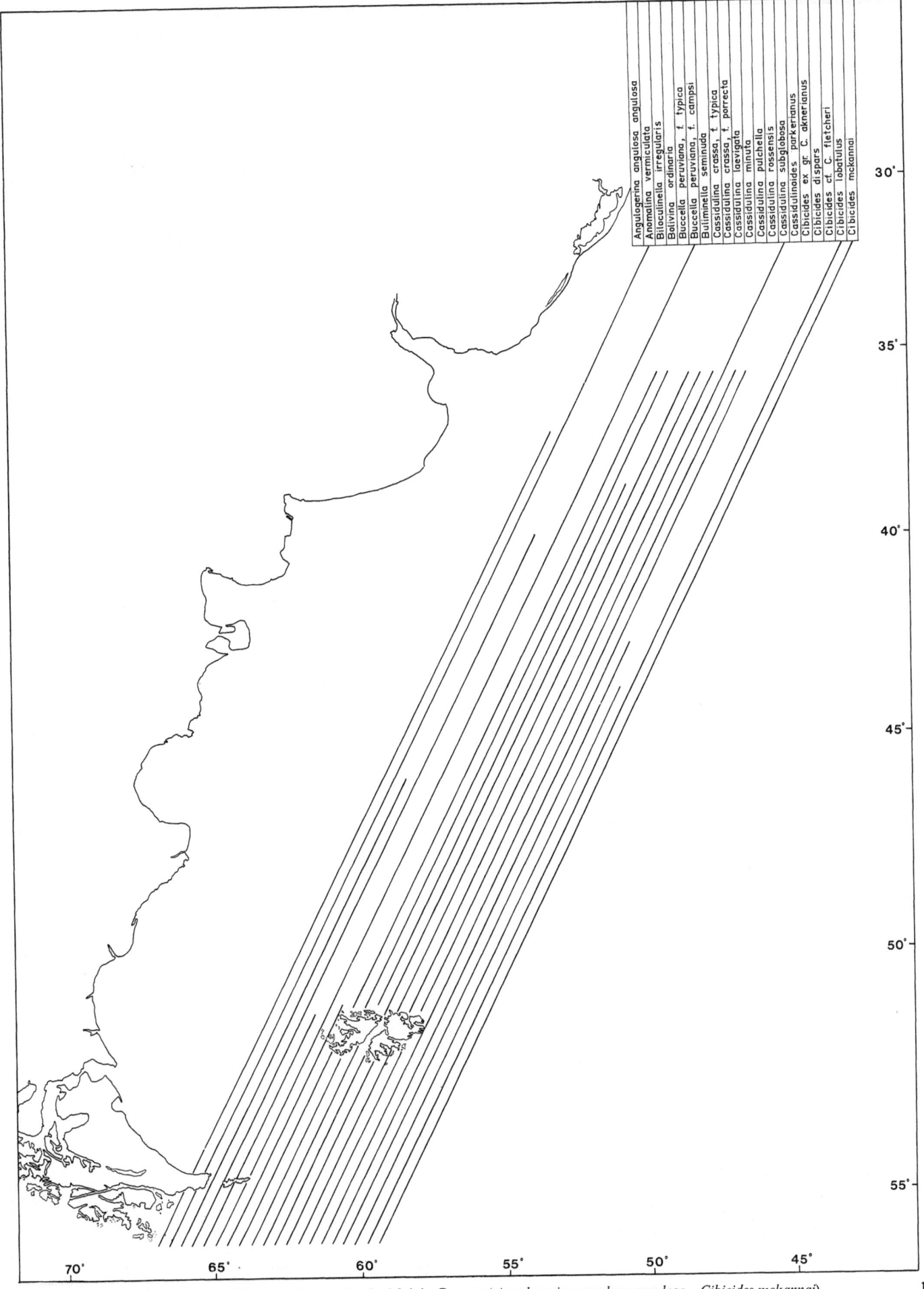

Fig. 13. Distribution of species in the Malvin Current (*Angulogerina angulosa angulosa – Cibicides mckannai*)

143

Labels (left to right):
Angulogerina angulosa angulosa
Anomalina vermiculata
Biloculinella irregularis
Bolivina ordinaria
Buccella peruviana, f. typica
Buccella peruviana, f. campsi
Buliminella seminuda
Cassidulina crassa, f. typica
Cassidulina crassa, f. porrecta
Cassidulina laevigata
Cassidulina minuta
Cassidulina pulchella
Cassidulina rossensis
Cassidulina subglobosa
Cassidulinoides parkerianus
Cibicides ex gr. C. aknerianus
Cibicides dispars
Cibicides cf. C. fletcheri
Cibicides lobatulus
Cibicides mckannai

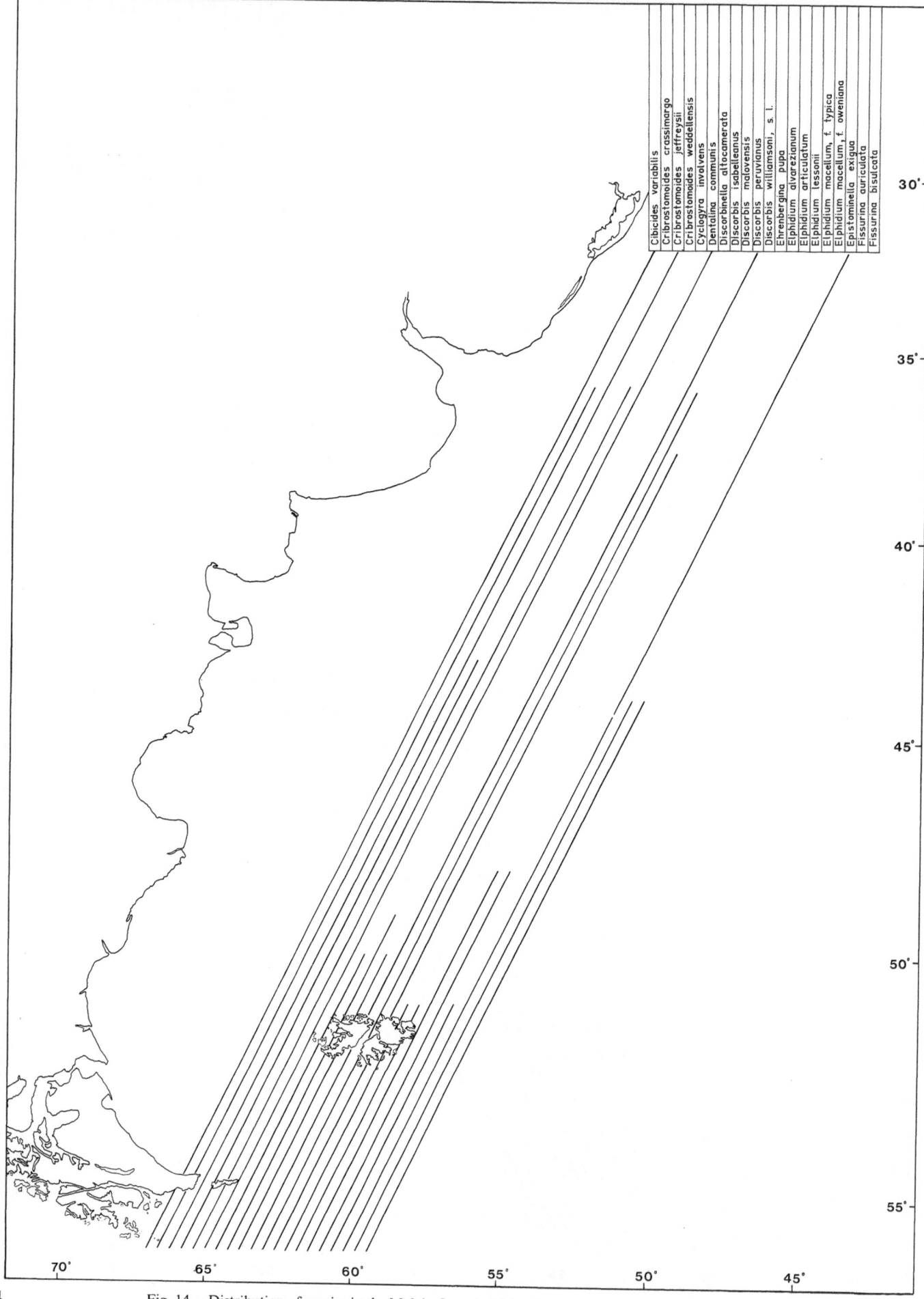

Cibicides variabilis
Cribrostomoides crassimargo
Cribrostomoides jeffreysii
Cribrostomoides weddellensis
Cyclogyra involvens
Dentalina communis
Discorbinella altocamerata
Discorbis isabelleanus
Discorbis malovensis
Discorbis peruvianus
Discorbis williamsoni, s. l.
Ehrenbergina pupa
Elphidium alvarezianum
Elphidium articulatum
Elphidium lessonii
Elphidium macellum, f. typica
Elphidium macellum, f. oweniana
Epistominella exigua
Fissurina auriculata
Fissurina bisulcata

Fig. 14.   Distribution of species in the Malvin Current (*Cibicides variabilis – Fissurina bisulcata*)

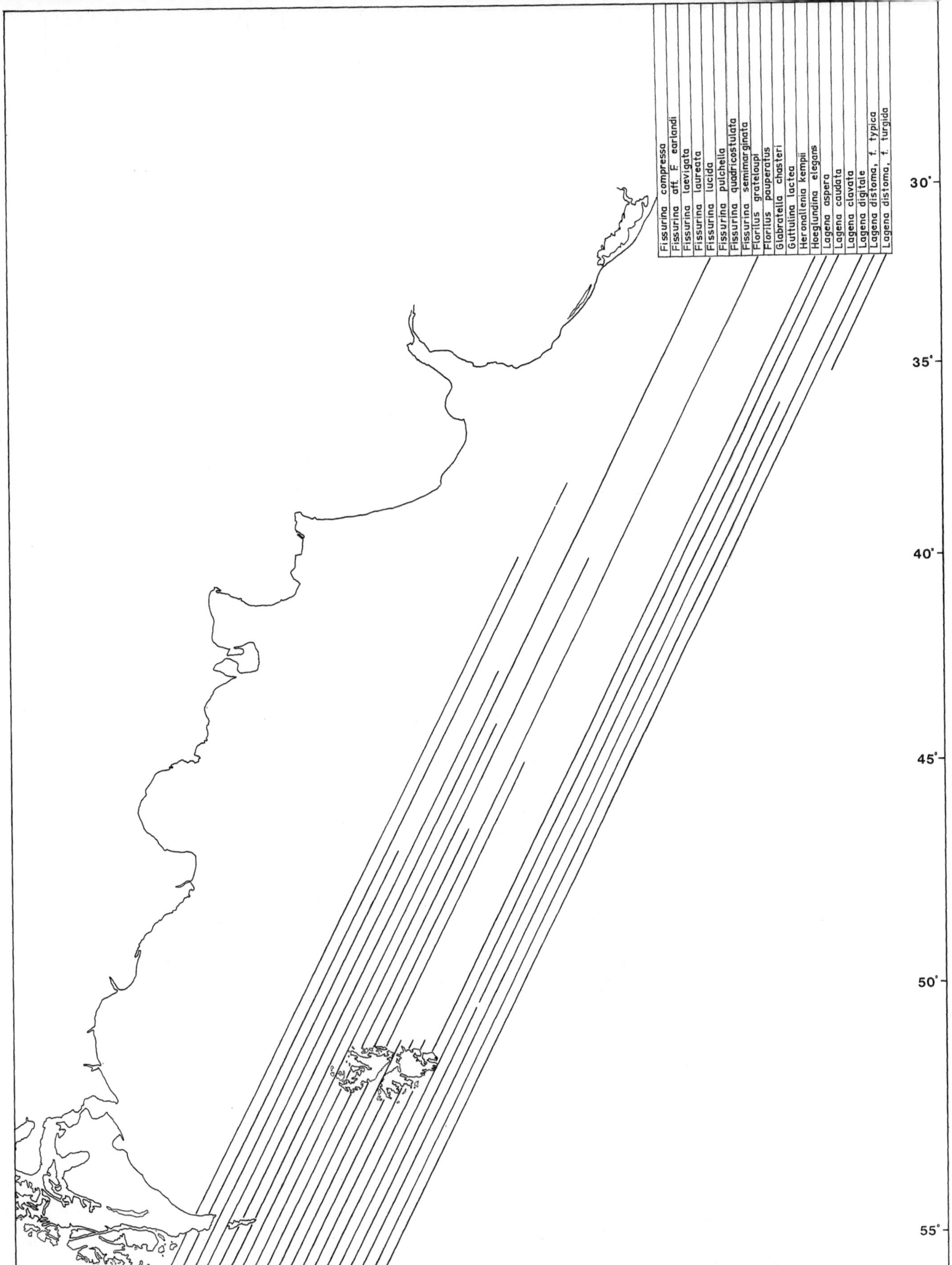

Fissurina compressa
Fissurina aff. F. earlandi
Fissurina laevigata
Fissurina laureata
Fissurina lucida
Fissurina pulchella
Fissurina quadricostulata
Fissurina semimarginata
Florilus grateloupi
Florilus pauperatus
Glabratella chasteri
Guttulina lactea
Heronallenia kempii
Hoeglundina elegans
Lagena aspera
Lagena caudata
Lagena clavata
Lagena digitale
Lagena distoma, f. typica
Lagena distoma, f. turgida
30°
35°
40°
45°
50°
55°
70°
65°
60°
55°
50°
45°

Fig. 15. Distribution of species in the Malvin Current (*Fissurina compressa – Lagena distoma*, f. turgida)

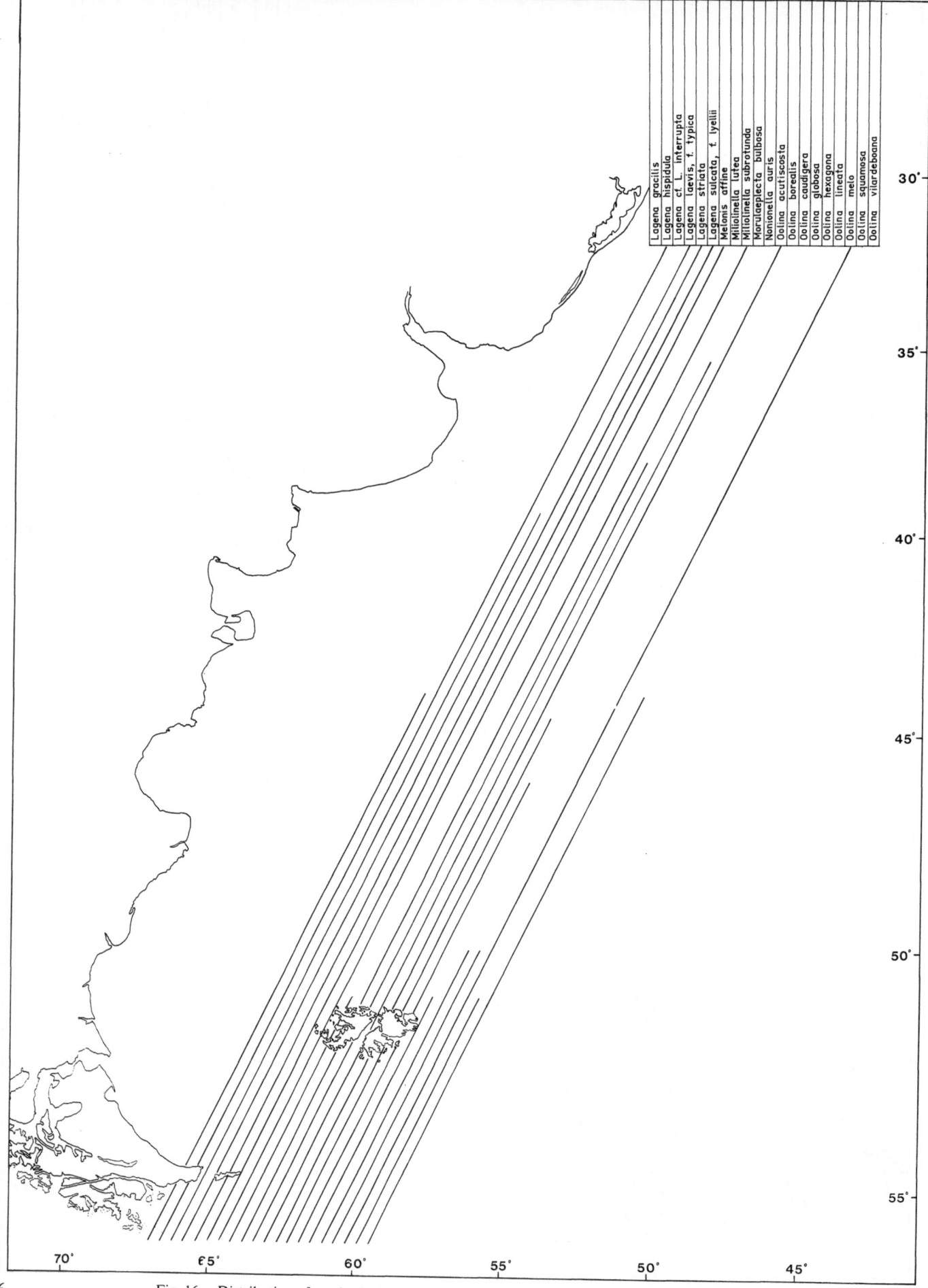

Lagena gracilis
Lagena hispidula
Lagena cf. L. interrupta
Lagena laevis, f. typica
Lagena striata
Lagena sulcata, f. lyellii
Melonis affine
Miliolinella lutea
Miliolinella subrotunda
Morulaeplecta bulbosa
Nonionella auris
Oolina acuticosta
Oolina borealis
Oolina caudigera
Oolina globosa
Oolina hexagona
Oolina lineata
Oolina melo
Oolina squamosa
Oolina vilardeboana

Fig. 16. Distribution of species in the Malvin Current (*Lagena gracilis – Oolina vilardeboana*)

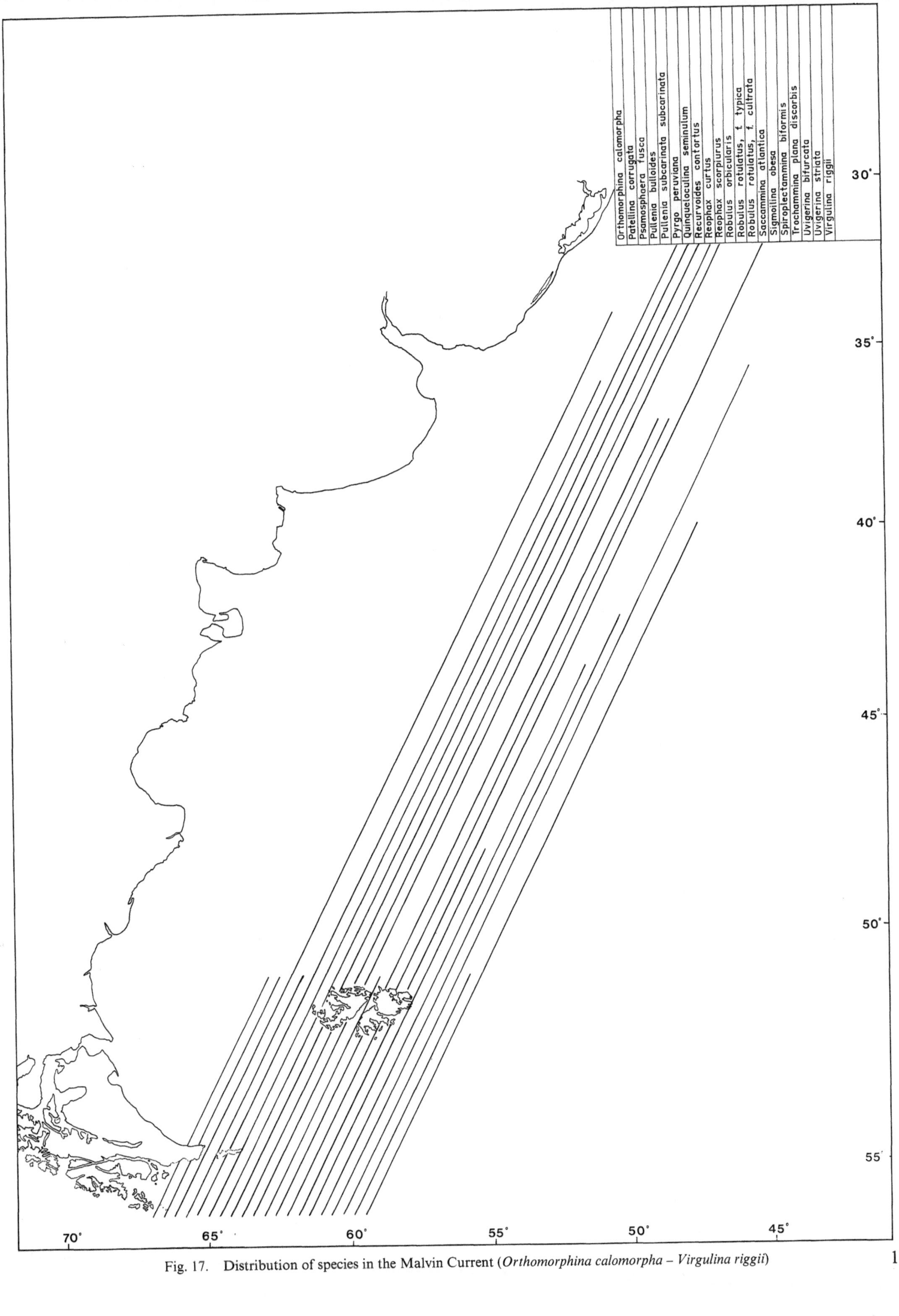
Orthomorphina calomorpha
Patellina corrugata
Psamosphaera fusca
Pullenia bulloides
Pullenia subcarinata subcarinata
Pyrgo peruviana
Quinqueloculina seminulum
Recurvoides contortus
Reophax curtus
Reophax scorpiurus
Robulus orbicularis
Robulus rotulatus, f. typica
Robulus rotulatus, f. cultrata
Saccammina atlantica
Sigmoilina obesa
Spiroplectammina biformis
Trochammina plana discorbis
Uvigerina bifurcata
Uvigerina striata
Virgulina riggii
30°
35°
40°
45°
50°
55°
70°
65°
60°
55°
50°
45°

Fig. 17. Distribution of species in the Malvin Current (*Orthomorphina calomorpha – Virgulina riggii*)